# Lecture Notes in Computer Science 2546

Edited by G. Goos, J. Hartmanis, and J. van Leeuwen

Springer
*Berlin*
*Heidelberg*
*New York*
*Barcelona*
*Hong Kong*
*London*
*Milan*
*Paris*
*Tokyo*

James Sterbenz   Osamu Takada
Christian Tschudin   Bernhard Plattner (Eds.)

# Active Networks

IFIP-TC6 4th International Working Conference, IWAN 2002
Zurich, Switzerland, December 4-6, 2002
Proceedings

Springer

Series Editors

Gerhard Goos, Karlsruhe University, Germany
Juris Hartmanis, Cornell University, NY, USA
Jan van Leeuwen, Utrecht University, The Netherlands

Volume Editors

James Sterbenz
BBN Technologies, Internetwork Research
10 Moulton Street 6/5C, Cambridge, MA 02138-1191, USA
E-mail: jpgs@ieee.org

Osamu Takada
Hitachi Ltd., Systems Development Laboratory
292 Yoshida-cho, Totsuka-ku, Yokohama-shi, Kanagawa-ken, 244-0817 Japan
E-mail: takada@sdl.hitachi.co.jp

Christian Tschudin
Uppsala University, Department of Information Technology
Box 337, 751 05 Uppsala, Sweden
E-mail: Christian.Tschudin@it.uu.se

Bernhard Plattner
The Swiss Federal Institute of Technology (ETH) Zurich
Computer Engineering and Networks Laboratory
Gloriastraße 35, 8092 Zurich, Switzerland
E-mail: plattner@tik.ee.ethz.ch

Cataloging-in-Publication Data applied for

A catalog record for this book is available from the Library of Congress.

Bibliographic information published by Die Deutsche Bibliothek
Die Deutsche Bibliothek lists this publication in the Deutsche Nationalbibliografie;
detailed bibliographic data is available in the Internet at <http://dnb.ddb.de>.

CR Subject Classification (1998): C.2, D.2, H.3.4-5, K.6, D.4.4, H.4.3

ISSN 0302-9743
ISBN 3-540-00223-5 Springer-Verlag Berlin Heidelberg New York

Springer-Verlag Berlin Heidelberg New York
a member of BertelsmannSpringer Science+Business Media GmbH

http://www.springer.de

© Springer-Verlag Berlin Heidelberg 2002
Printed in Germany

Typesetting: Camera-ready by author, data conversion by Boller Mediendesign
Printed on acid-free paper       SPIN: 10871657       06/3142       5 4 3 2 1 0

# Preface

This year marks the fourth year of the International Working Conference on Active Networks; this volume contains the papers accepted for full presentation at the conference. These proceedings are proof that Active Networks is a vibrant field with truly international participation, with research done not only in the context of formal programs (such as FAIN and DARPA Active Nets), but also in a number of independent research programs.

We received significantly more submissions than could possibly be accommodated during the two days of sessions, and combined with the desire for high standards this has resulted in a very selective program. The distinguished program committee did an outstanding job of reviewing the papers, with detailed comments and suggestions to the authors, each paper receiving from 4 to 6 reviews. Of the 53 papers submitted, 9 were accepted unconditionally. An additional set of papers were conditionally accepted, with program committee members working with the authors to insure that review concerns were addressed, while shepherding towards final publication; 11 of these papers are included in the proceedings. An additional set of poster presentations were given for worthy submissions that could not be accommodated within these limitations, as well as position papers and work in progress.

The best paper award went to Nadia Shalaby, Yitzchak Gottlieb, and Mike Wawrzoniak for "Snow on Silk: A NodeOS in the Linux Kernel," which begins these proceedings and the papers in the node operating systems session. These proceedings also contain active networking papers on: service deployment, discovery, and composition; monitoring and management; mobile wireless; system architecture; and peer-to-peer and group communication. In keeping with the "working" nature of the conference, significant time was reserved for discussion, and panels on active mobile networking, peer-to-peer, and the future prospects for active network research and deployment.

We would like to thank the technical program committee members for their hard work in reviewing and shepherding papers. We would like to thank Bernhard Plattner, Placi Flury, and Lukas Ruf for their considerable time and support in conference organization, and for providing the infrastructure at the Swiss Federal Institute of Technology (ETH), Zürich that made the work of the program committee considerably easier. And of course it is the work of the authors that is the core of these proceedings, and the attendees' participation that makes for an outstanding working conference. We hope you enjoy the fruits of their labor.

September 2002                                      James Sterbenz
                                                   Osamu Takada
                                                   Christian Tschudin

# Introduction

It is my great pleasure to introduce to you the proceedings of the 4th Annual International Working Conference on Active Networks, which took place in December 2002 at ETH Zürich, Switzerland. The end of 2002 has a special meaning for researchers in the area of active and programmable networks. It marks the closing of the DARPA Active Networks program, which provided important resources to researchers exploring this new approach to computer networking. This program, together with notable independent research in many other countries, generated a basic understanding of the architecture and functionality of active networks and produced a terminology which has been adopted by most individuals working in this field – terms like *NodeOS*, *Execution Environment*, or *Capsule* now have a well-known meaning. Furthermore, many potential applications of active networks have been studied, simulated, implemented, and demonstrated all over the world.

However, it is too early to declare victory. Active network technology still has not found widespread acceptance among manufacturers or network operators. The technology, in its "natural" shape, is disruptive, since introducing it would require that much of the current Internet would have to be changed. History teaches us that introducing a disruptive technology (prominent examples of such technologies are the personal computer and the Internet) takes a long time, if it replaces a well-established and well-working technology. Therefore, the active networks research community needs to consciously address the issues that may hinder a more widespread adoption of active networking. First, we have to make a better case and show that active network technology dominates competing technologies in real-life large-scale situations; this calls for a testbed which will not only be used for demonstrations, but also for production in our daily business. Second, we need to address the issue of how to introduce active network technology into our current IT environment, which today is largely dominated by the Internet – i.e., we need to take away some of the disruptiveness associated with active networks. Third, active networks research needs to look at and perhaps merge with related research areas; the "merger" with the area of programmable networks is imminent, but perhaps we also need to follow the road towards a software engineering discipline for active networks, which already has been paved by the research groups looking at domain-specific programming languages. Other interesting research areas are peer-to-peer networks and ad hoc networks, where nodes take on the roles of both end-systems and routers, and therefore may use active network technology to provide better and more suitable services to their users. Fourth, we need to be aware of basic technologies that may serve as drivers, such as network processors and programmable and reconfigurable hardware. Finally, we need to actively address the issue of technology transfer, i.e., to make active network technology accessible to manufacturers and operators.

IWAN 2002, coming at the end of the first phase of active network research, also marks the beginning of the second (and decisive) phase in our field. You will find some of the issues mentioned above addressed in the papers of this volume, but not all of them. Thus, there's still much to be done. Let's do it!

September 2002                                              Bernhard Plattner

# Organization

IWAN 2002 was organized by the Computer Engineering and Networks Laboratory, Swiss Federal Institute of Technology (ETH), Zürich. We would like to acknowledge the support of our sponsors, the Swiss Federal Institute of Technology, Zürich, IFIP (International Federation for Information Processing), Hitachi, Japan, and Fraunhofer FOKUS (Research Institute for Open Communication Systems), and we thank them for their contributions. Their support demonstrates the international interest in the benefits of active networking. The IWAN international committee would also like to express their gratitude to the Swiss Federal Office for Education and Science for their generous support.

## General Chair

Bernhard Plattner, Swiss Federal Institute of Technology (ETH), Zürich

## General Co-chair

Tadanobu Okada, NTT

## Program Co-chairs

James Sterbenz, BBN Technologies
Osamu Takada, Hitachi Ltd.
Christian Tschudin, Uppsala University

## Local Arrangements Chair

Lukas Ruf, Swiss Federal Institute of Technology (ETH), Zürich

## Local Arrangements Committee

Matthias Bossardt, Swiss Federal Institute of Technology (ETH), Zürich
Placi Flury, Swiss Federal Institute of Technology (ETH), Zürich
Ralph Keller, Swiss Federal Institute of Technology (ETH), Zürich

## Proceedings Production

Placi Flury, Swiss Federal Institute of Technology (ETH), Zürich
Lukas Ruf, Swiss Federal Institute of Technology (ETH), Zürich

## Treasurer

Rolf Wohlgemuth, Siemens Schweiz AG

## Steering Committee

Michitaka Kosaka, Hitachi Ltd.
Eckhard Moeller, Fraunhofer FOKUS
Olli Martikainen, IFIP TC6 WG6.7
Jonathan Smith, University of Pennsylvania
Radu Popescu-Zeletin, Technical University Berlin

# Technical Program Committee

Bobby Bhattacharjee, University of Maryland
Torsten Braun, University of Bern
Ken Calvert, University of Kentucky
Georg Carle, Fraunhofer FOKUS
Thomas Chen, Southern Methodist University
Hermann DeMeer, University College London
Andrzej Duda, INP Grenoble
Takashi Egawa, NEC Corporation
Ted Faber, USC/ISI
Mike Fisher, BTexact Technologies
Michael Fry, UTS
Alex Galis, University College London
Anastasius Gavras, Eurescom
Michael Hicks, University of Maryland, College Park
Gisli Hjalmtysson, Reykjavik University
David Hutchison, Lancaster University
Alden Jackson, BBN Technologies, Cambridge MA
Andreas Kind, IBM Zürich Research Laboratory
Yoshiaki Kiriha, NEC Corporation
Fumito Kubota, CRL
Tal Lavian, Nortel Networks Labs
Laurent Lefevre, INRIA/RESO/LIP
John Lockwood, Washington University
Thomas Magedanz, Fraunhofer FOKUS/TU Berlin
Ian Marshall, British Telecom
Doug Maughan, DARPA
Gary Minden, University of Kansas
Eckhard Moeller, Fraunhofer FOKUS
Masayuki Murata, Osaka University
Sandra Murphy, Network Associates Laboratories
Scott Nettles, University of Texas, Austin
Gian Picco, Politecnico di Milano
Bernhard Plattner, Swiss Federal Institute of Technology (ETH), Zürich
Nadia Shalaby, Princeton University
Yuvall Shavitt, Tel Aviv University
Jonathan Smith, University of Pennsylvania
Rolf Stadler, KTH Stockholm
Naoki Wakamiya, Osaka University
Tilman Wolf, University of Massachusetts
Miki Yamamoto, Osaka University
Hiroshi Yasuda, University of Tokyo
Martina Zitterbart, University of Karlsruhe

# Reviewers

C. Adam
K. Anagnostakis
D. Bauer
T. Becker
B. Bhattacharjee
T. Braun
C. Brou
K. Calvert
G. Carle
T. Chen
H. DeMeer
A. Duda
T. Egawa
T. Faber
M. Fisher
M. Fry
A. Galis
A. Gavras
R. Gold
R. Haas

M. Hicks
G. Hjalmtysson
D. Hutchison
S. Ioannidis
A. Jackson
S. Karnouskos
A. Kind
Y. Kiriha
F. Kubota
T. Lavian
K.-S. Lim
J. Lockwood
T. Magedanz
I. Marshall
D. Maughan
G. Minden
E. Moeller
J. Moore
M. Murata
S. Murphy

S. Nettles
G. Picco
B. Plattner
R. Pletka
M. Schoeller
N. Shalaby
Y. Shavitt
J. Smith
M. Solarski
R. Stadler
J. Sterbenz
O. Takada
C. Tschudin
N. Wakamiya
M. Waldvogel
T. Wolf
M. Yamamoto
M. Zitterbart

# Table of Contents

# Snow on Silk: A NodeOS in the Linux Kernel

Nadia Shalaby, Yitzchak Gottlieb, Mike Wawrzoniak, and Larry Peterson

Princeton University, Princeton NJ 08544, USA
{nadia,zuki,mhw,llp}@cs.princeton.edu

**Abstract.** Transferring active networking technology from the research arena to everyday deployment on desktop and edge router nodes, requires a NodeOS design that simultaneously meets three goals: (1) be embedded within a wide-spread, open source operating system; (2) allow non-active applications and regular operating system operation to proceed in a regular manner, unhindered by the active networking component; (3) offer performance competitive with that of networking stacks of general purpose operating systems. Previous NodeOS systems, Bowman, Janos, AMP and Scout, only partially addressed these goals. Our contribution lies in the design and implementation of such a system, a NodeOS within the Linux kernel, and the demonstration of competitive performance for medium and larger packet sizes. We also illustrate how such a design easily renders to the deployment of other networking architectures, such as peer–to–peer networks and extensible routers.

## 1 Introduction

A general architecture for active networks has evolved over the last few years [7, 25]. This architecture stipulates a three-layer stack on each active node, depicted in Figure 1. At the lowest layer, an underlying operating system (NodeOS) multiplexes the node's communication, memory, and computational resources among the various packet flows that traverse the node. At the next layer, one or more execution environments (EE) define a particular programming model for writing active applications. To date, several EEs have been defined, including ANTS [29, 28], PLAN [2, 11], SNAP [16], CANES [6] and ASP [5]. At the topmost layer are the active applications (AA) themselves. As evident from Figure 1, although more cumbersome for the user, an AA may access the NodeOS directly, forgoing EEs. From the application level perspective, such AAs may also be regarded as single operation EEs.

In the realm of an active network, security risks are heightened at the level of end-to-end active users, the active router node itself, the EEs and the active code that traverses the network, thereby necessitating a security component at each of these layers [1, 18]. Within an active node, the NodeOS necessarily maintains security on the OS level, depicted as a mandatory security module. EEs and AAs may also choose to maintain their own security mechanisms, depicted as optional security modules.

J. Sterbenz et al. (Eds.): IWAN 2002, LNCS 2546, pp. 1–19, 2002.

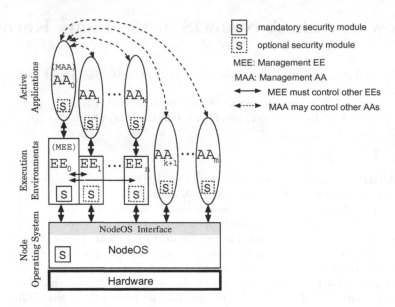

**Fig. 1.** The architecture of an active node: a Node Operating System above the Hardware exports a NodeOS Interface; Execution Environments define a programming model for Active Applications. EEs and AAs access the operating systems via the NodeOS Interface. With respect to the network as a whole, a Management AA and a Management EE are the ones bootstrapped at systems initialization, and manage AAs and EEs respectively. The NodeOS and the MEE necessarily maintain security on the node, while the MAA as well as EEs and AAs may implement their own security modules as an option.

Lastly, network control, monitoring and management is maintained by a Management EE (MEE), the first EE started at system initialization, which necessarily loads, manages and coordinates between subsequent EEs in the system, and maintains a mandatory security module. SENCOMM [12] is a comprehensive example of such a MEE. It is also possible for a Management AA (MAA) to exist, to similarly manage and coordinate between the AAs running on the node. If such a MAA acts with a role, or forgoes EEs to control AAs directly accessing the NodeOS, it would also maintain its own security module.

In order to transfer this powerful and flexible architecture from the research arena to wide scale everyday deployment, we need to design a NodeOS whose active networking functionality can be latent within a general purpose OS but, when necessary, implements this richer functionality without compromising performance. Therefore, our NodeOS design should simultaneously meet three goals: (1) be embedded within a wide-spread, open source operating system; (2) allow non-active applications and regular operating system operation to proceed in a regular manner, unhindered by the active networking component; (3) offer performance competitive with that of networking stacks of general purpose operating systems.

Previous NodeOS systems Bowman [15], Janos [26], AMP [8] and Scout [19], only partially addressed these goals. Our central contribution lies in the design and implementation of such a system, a NodeOS within the Linux kernel, and the demonstration of how we maintain performance comparable to Linux for routing on behalf of active applications. Besides contributing to decoupling the NodeOS specification from EEs/AAs from above, and the NodeOS implementation from below, we deliver a system with distinctly established protection boundaries between the AA/EE and NodeOS space, in terms of memory management, process scheduling and security. The benefits of our design are furthermore strengthened by the fact that it offers a straightforward mapping to other networking architectures, such as peer–to–peer networks and extensible routers, which would in turn lead to their deployment.

The remainder of this paper is organized as follows. The next section describes the design rationale and the recent key evolution of the NodeOS interface. Section 3 presents our architecture, which consists of Scout paths, their embedding into the Linux kernel as Silk, connecting to devices via the partner interface, and the functionality of two key device drivers communicating via the partner interface: Geni, a driver for Scout paths, and Sockets, a driver for user space. Section 4 describes the NodeOS implementation, how Snow and Sockets were used to interface to NodeOS API and user space respectively, and a brief description of the structure of the NodeOS module in Silk. We evaluate our design and implementation in Section 5, and demonstrate how it can be extended to conform to other networking paradigms in Section 6. Section 7 summarizes the related work to date and Section 8 concludes the paper.

## 2   NodeOS Interface

While each EE exports its own interface to the AAs, establishing a unified interface to the NodeOS is an essential element to the Active Networks architecture. We have contributed to the design and recent evolution of the NodeOS API specification [3], which has markedly influenced our NodeOS architecture and implementation. We briefly describe the primary abstractions of the NodeOS interface, outline the two kinds of data structures used, and highlight the recent evolution of the interface, which rendered our architectural design possible — in terms of the nature of data structure definitions, arguments to the API function calls, and the two models for memory allocation and thread management.

### 2.1   Abstractions

We define four primary abstractions: thread pools, memory pools, channels and files, encapsulating the node's computation, memory, communication and persistent storage, respectively; and a fifth abstraction, the domain, aggregates control and scheduling for the other four abstractions. Semantically, however, a particular domain does not necessarily subsume the instantiations of each of the other abstractions, or all of their components. Rather, its relationship to the other abstractions is summarized in Table 1, where we observe many-to-one, one-to-one and one-to-many mappings.

$$\text{Domain} \xrightarrow{M:1} \text{Memory Pool}$$
$$\text{Domain} \xrightarrow{1:1} \text{Thread Pool}$$
$$\text{Domain} \xrightarrow{1:M} \text{Channel}$$
$$\text{Domain} \xrightarrow{M:1} \text{File Name Space}$$

**Table 1.** Relationship of Domains to the Memory, Thread, Channel and Persistent Storage Abstractions

Channels are further characterized as either *incoming*, *inChan*s, characterized by a demultiplexing key specifying the chain of protocols, a buffer pool for packet queueing, and a function to handle the packets; or *outgoing*, *outChan*s, characterized by a processing key specifying a chain of protocols and the link bandwidth it's allowed. Moreover, *cut-through* channels, *cutChan*s, both receive and transmit packets, and are characterized by the *inChan/outChan* pair since they can also be constructed by concatenating the two. Cut-through channels are primarily motivated by the desire to allow the NodeOS to forward packets without EE or AA involvement.

This abstraction set allows us to view channels and domains as collectively supporting a flow-centric model: the domain encapsulates the resources that are applied to a flow, while the channel specifies what packets belong to the flow and what function is to be applied to the flow. The packets that belong to the flow are specified with a combination of addressing information and demultiplexing key, while the function that is to be applied to the flow is specified with a combination of module names and the handler function.

Other, non-primary abstractions include *events* scheduled by a domain, a *heap* for memory management, *packets* to encapsulate the data that traverses a channel, and the notion of *time* which NodeOS provides to EEs and AAs.

## 2.2   Objects and Specifications

The two major data structures of the API, are *specifications*, with visible, well-defined fields, and *objects*, which are opaque structures. The most important distinction being that, from the perspective of the NodeOS, the lifetime of a specification is a single API call, while the lifetime of an object is from its explicit creation via an API "create" call, until its explicit destruction, via an API "destroy" call.

For example, when creating a domain object via **an_domainCreate**, its an_-Domain argument is an opaque object, whose memory cannot be reused until the domain is destroyed; whereas its an_ThreadPoolSpec argument is a specification, whose memory, can be reused immediately after the **an_domainCreate** function returns.

As a result of these semantics, objects can be subcomponents of a specification, but not vice versa. From the EE's perspective, it must ensure that the specification and any storage it references is not mutated or reclaimed until the NodeOS API function returns. This enables the EE to pass the same specification to multiple APIs calls simultaneously.

From the perspective of the NodeOS, when a reference to a specification is passed as a parameter to a NodeOS API function, the NodeOS may freely access the specification and any storage it references, however it cannot modify it in any way. This guarantees that when the NodeOS API function returns, the EE will be able to locate pointers to storage that it placed into the specification.

We have intentionally defined both specifications and objects as pointer types in a uniform way. The distinction lies not in *how* they are defined, but rather in *where* they are defined. Specification structures need to be defined in the EE, and are therefore explicitly defined in the NodeOS API document [3]. Object structures, on the other hand, are defined in the NodeOS, since they need to remain implementation dependent, and as such, opaque to the API.

## 2.3   Arguments to API Calls

Arguments to API call functions can be standard C types, such as void *, char *, pointers to functions, such as void *(\*f)*(void * *arg*), objects, such as an_Domain, or specifications, such as an_ThreadPoolSpec.

Our decision to define both object and specification types as pointers allow us to pass them both by reference, as pointer type arguments. However, in the case of objects, the EE declares a pointer type of the form

```
typedef struct an_GenericObject * an_GenericTypeObject
```

in order to be able to make API calls with objects as arguments.

This recent change in the NodeOS API design provides three advantages. First, it allows for a set of portable header files, shared among various EEs. Second, this preserves the opaqueness of the objects from the EE's standpoint, thus allowing for different implementations of the NodeOS. And finally, as certain object types, such as the security credentials, become better understood, they could be migrated into specifications without changing the API (by simply adding their structure definitions to the EEs' common header files), thus preserving backward compatibility with all existing EEs and AAs.

## 2.4   Memory and Thread Management

Another key change was introducing *implicit* and *explicit* models for memory allocation and thread management. The former, and predictably more common case, delegates both to the NodeOS. The implicitly allocated objects will be automatically freed by the NodeOS when the corresponding destroy call is made, and all associated threads are also automatically destroyed.

The second, explicit model, serves the so called "trusted EEs", where the EEs and NodeOS are tightly coupled, sharing security and protection boundaries. In this case, the EE may wish to explicitly manage the memory used by the NodeOS for EE–created objects, and allocate its threads. It thus passes an appropriately–sized chunk of memory to all object create calls for the NodeOS to use. When such an explicitly allocated object is destroyed, the NodeOS tears down any internal state associated with the object and stored in the object memory, but does not free the memory.

This approach to the NodeOS specification allows for both trusted and untrusted EE implementations and gives the desired freedom in NodeOS design.

## 2.5   An API Example

The NodeOS API is specified via a portable set of header files, in C or Java, consisting of type definitions for objects, specifications and its API functions. E.g. the header file includes the following definition for incoming channel creation

an_InChan **an_inchanCreate**(void * *mem*, an_Domain *d*, an_DemuxKey *dmxKey*, char
   *protspec*, char *addrspec*, an_NetSpec *netspec*, an_ChanRecvFunc *deliverfunc*, void *
   *deliverarg* );

as well as for the an_NetSpec specification

```
typedef struct an_NetSpec {
        int             maxthreads;
        unsigned int    bandwidth;
        int      npbufs;
        an_PacketBuffer * pbufs;
} * an_NetSpec_t;
```

Additionally, EEs (or AAs directly) declare the opaque objects in a separate, non-portable header file, since they are only fully visible within the NodeOS, such as

```
typedef struct an_Domain * an_Domain_t;
```

# 3   Architecture

Our philosophy is that the provision of a modular networking tool-kit, coupled with good interface design to different system components, results in easy and efficient development of various networking architectures and services. The development essentially becomes a straightforward translation of one interface to another.

At the core of our architecture, is the observation that Scout paths [17] closely resemble the encapsulation of the primary abstractions of the NodeOS Interface. This observation has inspired the earlier work of implementing a NodeOS within the Scout stand–alone OS [19]. Moreover, architecting NodeOS around Scout paths results in employing the same mechanism for both the traditional and active forwarding services, thereby meeting our design goal of integrating the NodeOS interface in a manner that does not negatively impact our ability to forward non-active packets.

Unlike the Scout NodeOS implementation [19] however, and in line with our goal of employing a general purpose and wide spread OS, we embed the Scout path abstraction into Linux, known as "Scout in Linux Kernel" (Silk). This necessarily establishes user/kernel space boundaries, thus requiring a mechanism to communicate between them, the Scout paths, and other I/O devices.

In what follows, we explain the architecture of each these components and how they fit together to provide a NodeOS that meets our stated goals.

### 3.1   NodeOS Abstractions to Scout Paths

Designed to encapsulate I/O flows, a *Scout path* [17] is a structured system activity that provides an explicit abstraction for early demultiplexing, early packet dropping when flow queues are full, resource accounting per flow, explicit flow scheduling, as well as extensibility in adding new protocols and constructing new network services. Each Scout path consists of a string of protocol modules, and encapsulates a particular flow of data, such as a single TCP connection.

We can therefore characterize a path by its sequence of modules, a demultiplexing key, and the resource limits placed on the path (such as queue lengths, CPU cycle share, bandwidth). As such, this closely maps to the information required by the NodeOS: a domain is a container for the necessary resources (channels, threads, and memory), while a channel is specified by giving the desired processing modules and demultiplexing keys. As a consequence, we are able to design a NodeOS module to perform a simple mapping between Scout path operations to domain, channel, thread, and memory operations.

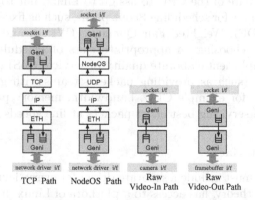

**Fig. 2.** Four Scout paths interfacing with user space via Sockets: a TCP path, a NodeOS path via UDP and two raw video paths, from a camera and onto a framebuffer

To demonstrate this concept, consider Figure 2, which portrays four examples of Scout paths. Both ends of a path are delineated by the Geni (Generic Interface)

module, which abstracts the device and communication interfaces to the path framework, performs demultiplexing, and maintains I/O queues at each end.

The first, non-NodeOS path, corresponds to a single TCP connection going through Ethernet, IP and TCP processing, respectively. From the bottom, packets arrive via the network interface, while user space applications connect to this path via the socket interface at the top. Transforming this scenario to one where, say UDP, packets are controlled via the NodeOS Interface, we simply chain a NodeOS module that performs the path–to–NodeOS–Interface functionality mapping above the UDP module. The result is the NodeOS path shown.

We can also construct *raw paths*, as depicted in Figure 2, where no protocol modules are chained to the path, resulting in very fast data forwarding. For example, in one case, packets are generated via a video camera, arrive on the camera interface, and are sent to the application via the socket interface. Geni abstracts the interfaces on each end, and maintains queues if buffering is needed. Likewise, the other raw path portrays an application that sends packets via the socket interface to be displayed via the framebuffer interface on video screen.

## 3.2   Silk

Conforming with our first design goal of using an established, open source OS, we chose to design our NodeOS within the Linux kernel, via Silk [4], which encapsulates the Scout path architecture into a loadable Linux module. In order to meet our second design goal, requiring that we preserve the ability to download and run standard non-active applications, Silk provides its own path scheduler and thread package that coexists with the Linux CPU scheduler.

The major advantage of permitting the coexistence of two OS schedulers, is that it enables the scheduler that is virtually in control, in our case, Silk, to not only decide what share of the CPU to assign to Linux, but to also maintain its own suite of schedulers for scheduling Scout paths, such as fixed priority, Earliest Deadline First (EDF), Weighted Fair Queueing (WFQ), and Best Effort Real Time (BERT). By choosing an appropriate class of schedulers, such as fixed priority, we can implement elaborate quality of service (QoS) processing via the NodeOS interface, (such as providing packet streams with guarantees on the system's resources, for example CPU, bandwidth, memory pools, threads and queues), while still servicing best effort packets at line speeds [20].

## 3.3   Partner Interface

Stand-alone Scout must include a separate driver for every device it uses. Silk on the other hand, in theory, has access to a plethora of Linux drivers. In practice, however, since Silk is really an embedded OS in its own right, a separate Scout module would be necessary to interface to every such driver. As the number of these devices increases, such a design becomes neither efficient nor scalable.

Instead, we designed a uniform and symmetric interface for intelligent devices, called the *partner interface* [27], applicable to all the heterogeneous devices within an OS. The key idea is that rather than rewrite the Silk drivers for all the Linux interfaces to file systems, devices (such as network cards), and various user level APIs (such as sockets), thus effectively redoing all the work done within the general OS for this specific embedded module, we unify all the drivers into the *partner interface*, where they all exchange a common data structure, the *partner vector*, and assume a symmetric *partner modus operandis* after OS initialization, as portrayed in Figure 3.

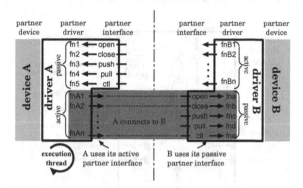

**Fig. 3.** The execution thread is at partner driver A. Therefore, partner driver A executes its active partner interface, thereby invoking the passive partner operations of device B.

The partner interface consists of five operations: *open*, *close*, *push*, *pull* and *control*; for opening and closing a device driver, sending and receiving data to and from the driver, and controlling the data flow, respectively. Partner operations exchange an *aggregate* internal data structure, the partner vector, or *pvec*, which is designed to simultaneously be an aggregate of one or more internal representations of any device driver (such as *iovecs*, *sk_buff*s, *frame_buffers*, etc.), and enable data manipulation among different device drivers without memory copying.

Each device driver must provide its side of the partner interface, thus translating its internal functionality to the partner operations. This allows communication flows established within a system to uniformly access the networking stack (e.g. via the Geni driver in the case of Scout paths), where data is forwarded or modified. Alternatively, if neither buffering nor additional processing is required, raw data can flow directly between a network interface card and an application via sockets due to the common partner interface.

Two particular partner drivers in Silk, Geni for Scout paths and Sockets for user space, are of special interest to the design of our NodeOS, and are therefore discussed in more detail below.

### 3.4 Geni: Scout Parnter Driver

From the partner interface standpoint within Silk, Scout paths are viewed as a device, and Geni, the driver to that device. Geni is conceptually architected to perform three major functions: (1) implement the partner operations, thereby translating partner operations to operations on Scout paths, such as *pathCreate*, pathDelete, demux, deliver and *drop*, and vice versa, respectively; (2) demultiplex the incoming packets, thus mapping network flows onto Scout paths (which the NodeOS modules maps to channels), and abstracting their forwarding functionality and (3) multiplex packets from the end of a particular path onto their target output devices, such as a network card or a socket.

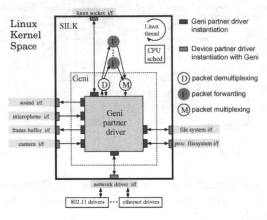

**Fig. 4.** Scout's partner interface, Geni, connects Scout paths to network and media devices, file systems and socket interfaces

A schematic picture of Geni within Silk is given in Figure 4, illustrating how Geni interfaces Scout paths to interact to the other drivers via the partner interface. Portrayed are the partner drivers of each interface from Geni's side and the other device's side, packet multiplexing and demultiplexing. The examples shown are network devices, file systems, video and audio devices, as well as the drivers that bridge the Linux kernel and user space, sockets.

Within Geni, the partner operations *open, close, push* and *pull* are roughly mapped to the Scout operations *pathCreate, pathDelete, deliver* and *dequeue*, respectively. As for the partner *ctl* operation, Geni translates it to the corresponding control operation on either a path, a particular Scout module, or on the Silk OS itself. Consequently, for active packets, the *ctl* translates to an operation on the NodeOS module.

Device partner interfaces are depicted as the small rectangles in Figure 4. Each such dark rectangle conceptually represents an instantiation of Geni partnered with a particular device driver. To illustrate this point, consider again the Scout paths depicted in Figure 2. For the TCP and NodeOS paths, the bottom and top Geni modules are instantiations of Geni partnering with the network and socket partners, respectively. In the case of raw paths, the bottom and top modules are instantiations of Geni's partnering with the video and socket partners, respectively.

### 3.5   Sockets: User Space Partner Driver

The standard UNIX socket interface bridges user space applications with kernel networking stacks. In our architecture, Silk hijacks the kernel networking stack from Linux, while still employing the socket driver (via the partner interface) to communicate with user space.

Like any partner driver, the socket partner driver translates the socket API functions, such as *socket, bind, listen, connect, read, write, select* and *ioctl*, to the five partner interface operations, and vice versa, also transforming the data from *iovec*s to *pvec*s and back. In compliance with our design goal of allowing non-active applications to run unhindered by the NodeOS functionality, the socket partner intercepts all Linux application calls that use standard networking protocols via the socket API, such as calls on the *PF_INET* family, and passes their processing on to Silk. This allows unmodified legacy applications to access TCP and UDP paths transparently. For a detailed description of the socket partner driver, the reader is referred to [27].

To handle active applications on the other hand, Silk provides a new, *PF_SCOUT* protocol family, which is registered with Linux at system initialization. The socket partner interfaces to Scout paths via Geni, as illustrated in Figure 2. Like the *PF_INET* family, the *PF_SCOUT* protocol family has to be exposed to user space, by the socket driver. It is noteworthy that via this mechanism, our design extends to experimental and non-standard protocols beyond active networking. All that is needed is implementing the corresponding protocol as a module chained into a Scout path.

## 4   Implementation

We assembled all the components of our architecture to implement the NodeOS functionality within Silk, and expose the NodeOS API to the EEs and AAs in Linux user space. The overall implementation in its entirety is depicted in Figure 5. In line with our design philosophy, the underlying mechanism of our NodeOS design is to provide a mapping from the existing system components and interfaces, to ones we architected and implemented so as to provide the active networking functionality. Specifically, a mapping from a user space AA or EE to an active path and vice versa occurs as follows:

| EE/AA | Snow | Linux kernel | Socket/Geni partner drivers | Silk |
|---|---|---|---|---|
| NodeOS API | ⟵⟶ Socket API | ⟵⟶ Linux Socket i/f | ⟵⟶ Partner Interface | ⟵⟶ Scout Path Operations |

To further elaborate on the implementation of the various components of our functional mapping in user space, Linux kernel to partner devices, and within a Scout path, we walk through an example of the creation of an incoming channel.

To create an incoming channel, an EE or AA calls the function

**an_inchanCreate(***mem, domain, dmxKey, protspec, addrspec, netspec, delfunc, delarg* **)**;

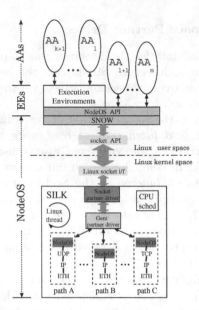

**Fig. 5.** Overall NodeOS implementation in Silk: NodeOS API is provided for user space; Snow translates it to the socket API calls; the Socket partner driver translates that to the partner operations, which communicate with Geni; Geni sets up and drives the active path in Silk, which includes a NodeOS module.

Since our architecture stipulates using the UNIX Socket API to cross the user/kernel boundary in Linux (making it consistent with non-NodeOS application calls), we implement a set of Silk NodeOS Wrappers (Snow) to convert the AA/EE calls written in the NodeOS API to the Socket APIs, depicted in Figure 5. Snow translates all NodeOS calls into *ioctl* calls on a socket file descriptor, of type *PF_SCOUT* family, with the corresponding NodeOS API function and arguments. Continuing with our example, to create a channel, Snow invokes

**fd** = **socket** (*PF_SCOUT, NOT_USED*, *pathTemplateNo*);
**ioctl(***fd*, *NODEOS_INCHAN_CREATE*, *argListPtr* );

Linux then converts these *socket/ioctl* calls at the user level to a sequence of kernel level calls. Namely, the *socket* call becomes a *sys_socket* call, which invokes *sock_create* to allocate memory for the socket structure, and subsequently invoking the *create* call, as follows:

**sys_socket(***PF_SCOUT, NOT_USED*, *pathTemplateNo* );
**sock_create(***PF_SCOUT, NOT_USED*, *pathTemplateNo, sock_str* );
**create(***sock_str*, *pathTemplateNo* );

The corresponding user space *ioctl* call, on the other hand, translates to the invocation of the following two kernel function calls in sequence

**sys_ioctl(***fd*, *NODEOS_INCHAN_CREATE*, *argListPtr* );
**ioctl(***inode*, *file*, *NODEOS_INCHAN_CREATE*, *argListPtr* );

After the *socket/ioctl* calls cross the user/kernel boundary in Linux, our socket partner driver needs to convert the respective Linux kernel function calls to the partner interface, so that it can then communicate with the corresponding Scout path via Geni. The socket *create* call is implemented by setting up a Silk socket structure, with corresponding operations and state. On the other hand, all *ioctl* calls are translated to the partner *ctl* operation. In our example, the socket partner driver invokes the following control operation on Geni as its partner

**ctl**(*Geni, NULL, NODEOS_INCHAN_CREATE, argListPtr, sizeof(argList)* );

For every control operation invoked on the Geni partner driver, Geni decides whether it is a control operation on a particular path, a module or on the Silk OS itself. In our case, it always translates to control operations on the NodeOS module. Specifically, Geni converts the above call to

**ctl**(*NodeOS, INCHAN_CREATE, argListPtr, sizeof(argList)* );

Once a call reaches the NodeOS module, it translates the NodeOS API functionality of setting up channels, maintaining domains, thread and memory pools into their corresponding operations on Scout paths. Thus, in our example, finally, the NodeOS module invokes *pathCreate* with the appropriate arguments to setup the Scout path.

# 5   Evaluation

Our experiments were conducted on 1.5GHz Pentium-4 machines, with a 256K cache, and 512MB RAM, connected to a 100Mbps network link via one to four 100Mbps Tulip cards. Each data point represents the average obtained for 10,000 packets, on three distinct runs.

To evaluate the robustness of our NodeOS router, we generated an increasing amount of packets on the two 100Mbps input ports, causing the router to forward to the other two 100Mbps output ports, to avoid input/output port contention. We were able to sustain line speed throughputs for packet sizes from 1 to 1400 bytes of payload, for three types of forwarding paths; (1) IP cut-through paths, representing the minimal router resources consumed to forward IP packets; (2) UDP cut-through paths, similar to null proxies, which forward UDP packets through Silk within the kernel; and (3) UDP active paths, where UDP packets cross into user space, and are processed via the NodeOS module in Silk. In other words, our router implementation coupled with the prototype hardware cannot be saturated with 200Mbps link speeds. This illustrates the robustness of our router and gives us ample resources for the additional functionality that we seek.

In our second experiment, we turn to latency measurements of the three aforementioned forwarding paths, which we depict for varying packet sizes in Figure 6. To put this evaluation in perspective, we include the latency component for a 100Mbps Ethernet hop per packet (half the latency cost is attributed to the sender and half to the receiver). We observe that latency linearly increases

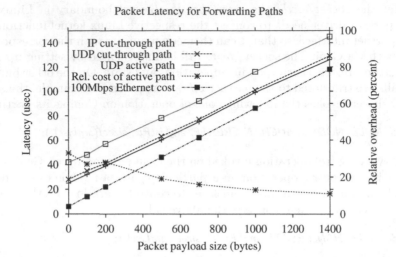

**Fig. 6.** Packet latency for three different forwarding paths (IP cut-through, UDP cut-through and UDP active paths), with the latency cost of one 100Mbps Ethernet hop.

with packet size. The UDP cut-through path is only marginally more costly than its minimal IP counterpart. The difference between the UDP cut-through and active paths is approximately 15 $\mu$sec, which quantifies the overhead of two components: crossing into user space and back from Silk (the bulk of this overhead), and the NodeOS Scout module (minimal portion of overhead). We also plot the relative cost of forwarding active packets through the router, which starts at 33% for minimal packet sizes, and gradually drops to 11% for maximum size packets.

To demonstrate how an active application with distinct data and control plane components makes use of our NodeOS architecture, we setup a third experiment that implements the wavelet dropper [9] on three machines — the source, intermediate, and sync nodes, representing the wavelet encoder, transcoder and decoder, respectively, as illustrated in Figure 7. The wavelet transcoder (or dropper) divides a wavelet encoded video stream into multiple layers. Depending on the level of congestion experienced at a router, high-frequency packet layers are dropped. Scout's RTP [22] protocol module records the number of packets successfully forwarded per flow, while the RTCP [22] protocol module uses this information to determine the available forwarding rate, and from this, the cutoff layer for forwarding, which it then sends back to the data forwarder.

We measured the latency for a wavelet video stream, encoded into maximum size Ethernet packets, from source (camera) to destination (video) via an intermediate transcoder node, with no network congestion, to be 306$\mu$sec. To assure this latency is independent of our particular implementation of the wavelet algorithm (as well as the EE it would be running on, in practice), we factored out

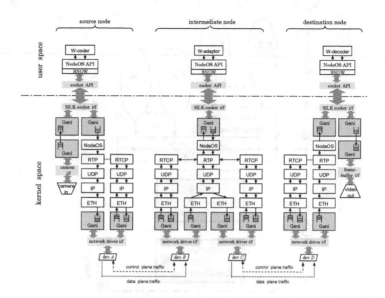

**Fig. 7.** Wavelet dropper application invoking RTP and RTCP paths from user space. Source, intermediate and sync nodes execute the wavelet encoder, adaptor and decoder, respectively.

the time spent in the user space wavelet encoder, transcoder and decoder modules, thereby only reporting the remaining relevant part of this configuration, including the two network hops, as well as the three roundtrip crossings from kernel to user space. The network latency accounted for 236$\mu$sec, or 77% of this cost, demonstrating that our NodeOS implementation, along with its ability to process packets at line speeds, was efficient in providing the extra RTP/RTCP functionality within the kernel, and crossing into user space for setting up the active video streams.

## 6  Design Extensibility

An important feature of our base design is that apart from providing a NodeOS that meets our design goals, it enables us to deploy other systems with different networking architectures. One example is that we can embed a hierarchical extensible router into our framework, as shown in Figure 8.

That particular extensible router [24] is comprised of connecting the Pentium, on which we run Silk, via the PCI bus to a network processor, namely an Intel IXP1200, which in turn consists of a StrongARM and a set of six MicroEngines executing in parallel. This results in four execution levels (shown in Figure 8), partitioning the router functionality among the hardware and software in concert, and corresponds to four distinct classes of paths through the system, from the fastest, at the lowest MicroEngine level, to the slowest, crossing into user

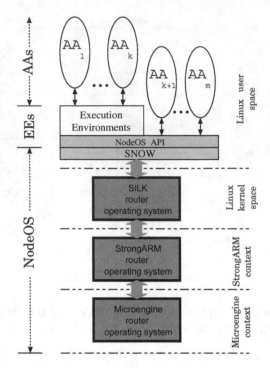

**Fig. 8.** Mapping our NodeOS architecture to that of an Extensible Router —
active packets make use of the entire router hierarchy for data and control flows.

space. From Silk's side, the PCI bus bridging to the IXP is accomplished via a
special partner driver, `vera.o` [14].

Under such a setup, we view the NodeOS as the OS of the distributed extensible router, roughly spanning the lower three execution levels of Figure 8.
Consequently, our design permits the NodeOS API abstractions to take advantage of the entire hierarchical OS of the prototype router, with its ability to
process packets at line speeds, despite the added extensibility and functionality.

In a similar fashion, we can employ our base architecture to design a peer–
to–peer (P2P) substrate node, such as Pastry [21]. For illustrations on employing
our architecture in implementing substrate nodes of two example peer–to–peer
systems, the reader is referred to a more detailed version of this paper [23].

## 7    Related Work

To date, several NodeOS implementations have been reported. Bowman [15],
runs on generic UNIX substrate, and therefore cannot provide fine grained resource control. Additionally, being the first NodeOS, it was developed in the
early days of the specification and provides only a subset of the interfaces.

Three other NodeOS implementations were discussed and compared in [19].
The first, implemented within stand–alone Scout [17], is the closest to our Silk

implementation, where channel fucntionality is wrapped around Scout paths. However, no protection domains were maintained, necessitating EEs and AAs to implement their own security. Another shortcoming is that Scout is not a commonly used OS. Relative to that, the main shortcoming of our NodeOS in Silk is the overhead of context switching for small packets, as we have seen in Section 5.

Janos [26] is based on the OSKit component base [10], where the NodeOS and EEs shared a protection domain — running as a kernel on top of hardware or as one user space process on top of a Unix OS. Such a single address space system cannot guarantee separation between concurrent EEs. Finally, AMP [8] is layered on top of the MIT exokernel [13], and the NodeOS was implemented within a user space OS library, with its own memory and thread management. Although this design comparatively taxed system performance, its primary focus on security contributed several issues to the ongoing design of the NodeOS interface.

# 8   Conclusions

We set out to design and build a NodeOS which would be embedded within a widely used OS, while allowing non-active applications and regular OS operation to proceed in a regular manner, unhindered by the active networking component, and at the same time offer performance competitive with that of networking stacks of general purpose operating systems.

Our central contribution was to simultaneously satisfy these design goals with our underlying design mechanism: use established system components (namely Linux, Scout paths and Linux device drivers) with well known interfaces (such as sockets and the module interface to Scout paths), and construct mappings between those and the APIs and operations we designed for NodeOS (namely the NodeOS API, Snow and the Partner Interface), thereby delivering kernel–level NodeOS functionality from user space.

Additionally, we have contributed to decoupling the NodeOS specification from EEs/AAs from above, and the NodeOS implementation from below, thereby delivering a *portable* set of header files, allowing EE/AA/NodeOS modifications in any direction. This resulted in the additional contribution of delivering a system with distinctly established protection boundaries between the AA/EE and NodeOS space, in terms of memory management, process scheduling and security.

We demonstrated that our system is robust, sustaining line speed bandwidth for cut-through and active flows that cross into user space, and measured that the relative overhead cost of forwarding active paths within our NodeOS is around 11% for maximal size packets. Forwarding latencies were shown to scale linearly with packet sizes for cut-through and active paths, ranging from $40\mu sec$ to $145\mu sec$ for minimal and maximum sized active packets, respectively. A prototype application that uses the NodeOS API, the wavelet dropper, exhibited regular and expectable performance.

An additional contribution is the extensibility of our architecture. We demonstrate how to use our base architecture to deploy a P2P system or an extensible router by only changing the user space API from above, or extending the NodeOS to a hierarchical system from below, respectively.

# References

[1] Active Networks Security Working Group. Security architecture for active nets. Available as ftp://www.ftp.tislabs.com/pub/anfr/secrarch5.ps, November 2001.

[2] D. S. Alexander, M. Shaw, S. M. Nettles, and J. M. Smith. Active Bridging. In *Proceedings of the ACM SIGCOMM '97 Conference*, pages 101–111, September 1997.

[3] AN NodeOS Working Group. NodeOS interface specification. Available as http://www.cs.princeton.edu/nsg/papers/nodeos.ps, January 2002.

[4] A. Bavier, T. Voigt, M. Wawrzoniak, and L. Peterson. SILK: Scout Paths in the Linux Kernel. Technical Report TR–2002–009,, Department of Information Technology, February 2002.

[5] S. Berson, B. Braden, T. Faber, and B. Lindell. The ASP EE: An Active Network Execution Environment. In *Proc. of the DARPA Active Networks Conference and Exposition (DANCE)*, San Francisco, CA, May 2002.

[6] S. Bhattacharjee, K. Calvert, and E. Zegura. Congestion control and caching in CANES. In *ICC '98*, 1998.

[7] K. Calvert. Architectural framework for active networks. Available as http://www.cs.gatech.edu/projects/canes/papers/arch1-0.ps.gz, July 1999.

[8] H. Dandekar, A. Purtell, and S. Schwab. AMP: Experiences in building an exokernel–based platform for active networking. In *Proc. of the DARPA Active Networks Conference and Exposition (DANCE)*, San Francisco, CA, May 2002.

[9] M. Dasen, G. Fankhauser, and B. Plattner. An Error Tolerant, Scalable Video Stream Encoding and Compression for Mobile Computing. In *Proceedings of ACTS Mobile Summit 96*, pages 762–771, November 1996.

[10] B. Ford, G. Back, G. Benson, J. Lepreau, A. Lin, and O. Shivers. The Flux OSKit: A substrate for OS and language research. In *Proceedings of the 16th ACM Symposium on Operating Systems Principles*, pages 38–51, St. Malo, France, October 1997.

[11] M. Hicks, P. Kakkar, J. T. Moore, C. A. Gunter, and S. Nettles. PLAN: A packet language for active networks. In *Proceedings of the 3rd ACM SIGPLAN International Conference on Functional Programming Languages*, pages 86–93, September 1998.

[12] A. W. Jackson, J. P. Sterbenz, M. N. Condell, and R. R. Hain. Active Monitoring and Control: The SENCOMM Architecture and Implementation. In *Proc. of the DARPA Active Networks Conference and Exposition (DANCE)*, San Francisco, CA, May 2002.

[13] M. F. Kaashoek, D. R. Engler, G. R. Ganger, H. Briceno, R. Hunt, D. Mazieres, T. Pinckney, R. Grimm, J. Jannotti, and K. Mackenzie. Application performance and flexibility on exokernel systems. In *Proceedings of the 16th ACM Symposium on Operating Systems Principles*, pages 52–65, St. Malo, France, October 1997.

[14] S. Karlin and L. Peterson. VERA: An Extensible Router Architecture. *Computer Networks*, 38(3):277–293, 2002.

[15] S. Merugu, S. Bhattacharjee, E. Zegura, and K. Calvert. BOWMAN: A node OS for Active Networks. In *Infocom 2000*, March 2000.

[16] J. T. Moore, M. Hicks, and S. Nettles. Practical programmable packets. In *Proceedings of the Twentieth IEEE Computer and Communication Society IN-FOCOM Conference*, pages 41–50, April 2001.

[17] D. Mosberger and L. L. Peterson. Making Paths Explicit in the Scout Operating System. In *Proceedings of the Second USENIX Symposium on* , pages 153–167, Seattle, WA USA, October 1996.

[18] S. Murphy, E. Lewis, R. Puga, R. Watson, and R. Yee. Strong security for active networks. In *Proceedings of the Open Architectures 2001 Conference*, pages 1–8, Anchorage, AK USA, April 2001.

[19] L. Peterson, Y. Gottlieb, M. Hibler, P. Tullmann, J. Lepreau, S. Schwab, H. Dandelkar, A. Purtell, and J. Hartman. An OS Interface for Active Routers. *IEEE Journal on Selected Areas in Communications*, 19(3):473–487, March 2001.

[20] X. Qie, A. Bavier, L. Peterson, and S. C. Karlin. Scheduling Computations on a Software-Based Router. In *Proceedings of the ACM SIGMETRICS 2001 Conference*, pages 13–24, June 2001.

[21] A. Rowstron and P. Druschel. Pastry: Scalable, distributed object location and routing for large-scale peer-to-peer systems. In *IFIP/ACM International Conference on Distributed Systems Platforms (Middleware)*, pages 329–350, November 2001.

[22] H. Schulzrinne, S. Casner, R. Frederick, and V. Jacobson. RTP: A transport protocol for real-time applications; RFC 1812. *Internet Request for Comments*, January 1996.

[23] N. Shalaby, Y. Gottlieb, and L. Peterson. Snow on Silk: A NodeOS in the linux kernel. Technical Report TR–641–02, Department of Computer Science, Princeton University, June 2002.

[24] N. Shalaby, L. Peterson, A. Bavier, Y. Gottlieb, S. Karlin, A. Nakao, X. Qie, T. Spalink, and M. Wawrzoniak. Extensible Routers for Active Networks. In *Proc. of the DARPA Active Networks Conference and Exposition (DANCE)*, pages 92–116, San Francisco, CA, May 2002.

[25] J. M. Smith, K. L. Calvert, S. L. Murphy, H. K. Orman, and L. L. Peterson. Activating Networks: A Progress Report. *IEEE Computer*, 32(4):32–41, April 1999.

[26] P. Tullmann, M. Hibler, and J. Lepreau. Janos: a Java–oriented OS for Active Networks. *IEEE Journal on Selected Areas in Communications*, 19(3), March 2001.

[27] M. Wawrzoniak, N. Shalaby, and L. Peterson. Intelligent Devices as Symmetric Partners: Architecture and Implementation. Technical Report TR–642–02, Department of Computer Science, Princeton University, January 2002.

[28] D. Wetherall. Active network vision and reality: lessons from a capsule-based system. In *Proceedings of the 17th ACM Symposium on Operating Systems Principles*, pages 64–79, December 1999.

[29] D. Wetherall, J. Guttag, and D. Tennenhouse. ANTS: A toolkit for building and dynamically deploying network protocols. In *IEEE OPENARCH 98*, San Francisco, CA, April 1998.

# PromethOS:
# A Dynamically Extensible Router Architecture Supporting Explicit Routing

Ralph Keller, Lukas Ruf, Amir Guindehi, Bernhard Plattner

Computer Engineering and Networks Laboratory
Swiss Federal Institute of Technology, Switzerland
{keller | ruf | guindehi | plattner}@tik.ee.ethz.ch

**Abstract.** Commercially available routers typically have a monolithic operating system that cannot be easily tailored and upgraded and support new network protocols. PromethOS is a modular router architecture based on Linux, which can be dynamically extended by plugin modules that are installed in the networking kernel. To install and configure plugins we present a novel signaling protocol that establishes explicitly routed paths transiting selected nodes in a predefined order. Such paths can be non simple, where a given node is being visited more than once.

**Keywords.** Active networking, extensible router architecture, explicit path routing, service deployment

## 1 Introduction

In the past, the functionality of routers was very limited, namely forwarding packets based on the destination address. Recently, new network protocols and extensions to existing protocols have been proposed and are being deployed, requiring new functionality in modern routers at an increasingly rapid pace. However, present day commercially available routers typically employ a monolithic architecture which is not easily upgradable and extensible to keep up with new innovations.

This paper presents the design and implementation of PromethOS,[1] an innovative router architecture with a modular design that can be extended to support new and dynamically deployed protocols. The specific objectives of this architecture are as follows:

- *Modularity.* The router architecture is designed in a modular fashion with components coming in form of plugins which are modules that are dynamically loaded into the kernel and have full kernel access without crossing address spaces.

---

[1] PromethOS originates from Prometheus who was the wisest Titan according to the Greek mythology. His name means „forethought" and he was able to foretell the future. The project was initially codenamed COBRA.

J. Sterbenz et al. (Eds.): IWAN 2002, LNCS 2546, pp. 20–31, 2002.
© Springer-Verlag Berlin Heidelberg 2002

- *Flexibility*: For each plugin class, multiple plugin instances can be created. Different configurations of the same plugin can co-exist simultaneously in the kernel, with plugin instances sharing the same code but operating on their own data.

- *Packet classification*: By defining filters, incoming data packets are classified to belong to a data flow and by binding a plugin instance to such a flow, all matching packets will be processed by the corresponding plugin instance.

- *Performance*: An efficient data path is guaranteed by implementing the complete data path in kernel preventing costly context switches.

- *Code Deployment*: Efficient mechanisms exist to retrieve plugins from remote code servers, install and configure them, and to setup network wide paths such that traffic transits these plugins as desired by the application.

- *Integration in Linux*: The implementation needs only minimal changes to the existing Linux source code and can easily be integrated into newer releases.

We have implemented our framework based on the Linux kernel. We have selected this platform because of its portability, freely available source code, extensive documentation, and wide spread use as a state of the art experimental platform by other research groups. Due to its modularity and extensibility, we are convinced that our proposed framework makes it a useful tool for researchers in the field of programmable router architectures and protocol design. All our code is released in the public domain and can be retrieved from our website [ ].

The main contributions of this paper are as follows:

- Design and implementation of a modular and extensible *node architecture* that allows code modules to be dynamically loaded into the networking subsystem at runtime.

- Design and implementation of a novel *signaling protocol* to establish explicitly routed paths through the network and the installation and configuration of plugins along such paths.

In the remainder of this paper, we discuss the design and implementation of our framework. In Section , we first focus on a single node, describe the architecture, and consider how it can be extended by installing plugins into the networking subsystem. We demonstrate an example use of the PromethOS plugin framework to give the reader a feel of how the architecture can be used. Section then focuses at the network scope, discusses how explicitly routed paths can be setup, how plugins are retrieved from remote code repositories and installed on selected nodes. Section reviews related work and Section concludes this paper.

## PromethOS Node Architecture

### Architectural Overview

The main objective of our proposed architecture is to build a modular and extensible networking subsystem that enables to deploy and configure packet processing components for specific flows. Figure illustrates our dynamically extensible router architecture.

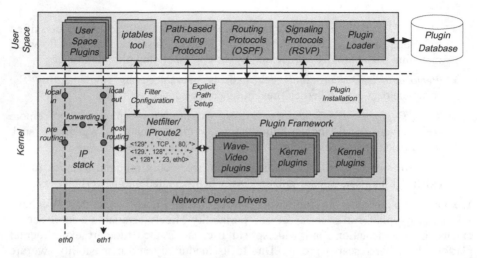

**Figure.** PromethOS modular and extensible router architecture

The most important components are as follows:

- *Network device drivers* implement hardware specific send and receive functions. Packets correctly received from an interface enter the IP stack.

- *Netfilter* classifies packets according to filter rules at various hooks. Packets matching a filter are passed to registered kernel modules for further processing.

- The *plugin framework* provides an environment for the dynamic loading of plugin classes, the creation of plugin instances as well as their configuration and execution.

- The *plugin loader* is responsible for requesting plugins from remote code servers which store plugin classes in a distributed *plugin database*.

- The *path based routing* protocol is used to setup explicitly routed paths and to install plugins on selected nodes.

- Other *routing and signaling* protocols compute the routing table and provide resource reservation mechanisms.

## Netfilter Framework

The netfilter framework [ ] provides flexible packet filtering mechanisms which are performed at various hooks inside the network stack. Kernel modules register callback functions that get invoked every time a packet passes the respective hook. The user space tool iptables allows to define rules that are evaluated at each hook. A packet that matches these rules is handed to the target kernel module for further processing. The netfilter framework together with the iptables tool provide the minimum mechanisms required to load modules into the kernel specifying packet matching rules evaluated at hooks and the invocation of the matching target module.

However netfilter has a serious restriction since all loadable modules must be *known at compile time* to guarantee proper kernel symbol resolution for the linking process. Thus only kernel modules that have been *statically configured* can be loaded into the networking subsystem. This is a significant limitation since we envision a router architecture that allows to load arbitrary *new* components at *runtime*.

### Plugin Framework and Execution Environment

To overcome this limitation we are extending the netfilter framework with a *plugin framework*. The plugin framework manages all loadable plugins and dispatches incoming packets to plugins according to matching filters. When a plugin initially gets loaded into the kernel it registers its virtual functions with the plugin framework. Once a packet arrives and needs to be processed by a plugin the framework invokes the previously registered plugin specific callback function. Since plugins register their entry points the entry functions do not need to be known at compile time and for this reason the plugin framework can load and link any plugin into the kernel.

Every PromethOS plugin offers an input and output channel in accordance with [ ] representing a *control* and *reporting* port. The control port is used for managing the PromethOS plugin such as configuration, the reporting port is read only to collect status information from the plugin.

### Plugin Classes and Instances

For the design of plugins we follow an object oriented approach. A *plugin class* is a dynamically loadable Linux kernel module that specifies the general behavior by defining how it is initialized configured and how packets need to be processed. A *plugin instance* is a runtime configuration of a plugin class bound to a specific flow. An instance is identified by a node unique instance identifier. In general it is desirable to have multiple configurations of a plugin each having its own data segment for internal state. Multiple plugin instances can be bound to one flow and multiple flows can be bound to a single instance. Through a virtual function table each plugin class responds to a standardized set of methods to initialize configure reconfigure itself and for processing packets. All code is encapsulated in the plugin itself, thus the plugin framework is not required to know anything about a plugin's internal details. Once a packet is associated with a plugin the plugin framework invokes the processing method of the corresponding plugin passing it the current instance data segment and a pointer to the kernel structure representing the packet struct sk_buff.

### Control from User Space

PromethOS and its plugins are managed at load time by providing configuration parameters and at run time through the control interfaces via the procfs. When the PromethOS plugin framework initially gets loaded it creates the entry

/proc/promethos. Below this entry the control and reporting ports of individual plugins are registered. PromethOS plugins are loaded by iptables which we extended with semantics required for the PromethOS plugin framework. The communication to control plugins and report messages between user space and plugins follows a request reply approach. A control message is addressed to the appropriate plugin by passing the plugin instance identifier as a parameter and the plugin then responds with a reply.

## Example Use of PromethOS Plugin Framework

To give the reader a feel for the simplicity and elegance with which plugins can be put into operation, we illustrate the commands necessary to load and configure a WaveVideo [ ] plugin performing video scaling. Note that these commands can be executed at any time, even when network traffic is transiting through the system. As mentioned above, we use a PromethOS enhanced iptables program that interacts with the iptables framework. In the extension of iptables, we implement calls to the insmod program which serves as the primary tool to install Linux kernel modules.

- Loading and registering plugin:

  iptables -t promethos -A PREROUTING -p UDP -s             -dport
  -j PROMETHOS   -plugin WV   -autoinstance   -config

  This command adds a filter specification to the PromethOS plugin framework, requesting to install the WV plugin at the PREROUTING hook and creating an instance of this plugin to perform video scaling at       Byte/s. If the plugin framework is not yet loaded, the module dependency resolution of Linux installs it on demand.

- Upon successful completion of the plugin loading and instantiation, the plugin framework reports the plugin instance number:
  Router plugin instance is

- By this instance number, the plugin control port can be accessed:
  echo            > /proc/promethos/net/management
  This reconfigures the WV plugin to scale the video to a maximum output of       Byte/s.

- The configuration of the PromethOS table can be retrieved with iptables:
  iptables -t promethos -L
  Chain PREROUTING (policy ACCEPT)
  target    prot opt source        destination
  PROMETHOS udp             anywhere udp dpt     WV

- The plugin and the framework may be removed from the kernel by the standard mechanisms provided by iptables and the Linux kernel module framework.

This example demonstrates the seamless integration of the PromethOS plugin framework in Linux, allowing to load arbitrary code at runtime.

## Code Deployment on Explicitly Routed Paths

In the previous section we have presented our active node architecture that can be dynamically extended with components coming in form of loadable kernel modules. The mechanisms illustrated for installing and configuring plugins require *local access* to the router which is a feasible approach for setting up routers with a static configuration. However the active networks paradigm envisions an infrastructure that can be programmed by network administrators and end users in a more flexible fashion. For active networks we need new routing mechanisms that take into account that end to end paths include processing sites.

In this section we present a novel signaling protocol that allows to *deploy plugins on selected nodes* and to *establish paths transiting these nodes* in a given order. In the context of active networks conventional destination based routing schemes cannot satisfy the requirements demanded by active applications since traffic needs to transit processing sites generally *not located* on the IP default path. In our opinion the introduction of new code into routers should be performed in a *structured way* where network service providers or end users *explicitly configure* the network with the required functionality enabling efficient allocation of network resources among competing applications.

Finding an optimal routing *and* processing path can be seen in the context of constraint based routing where processing constraints define requirements on the order and location of processing functions. A suitable algorithm that finds an optimal route through a sequence of processing sites has been proposed in [ ].

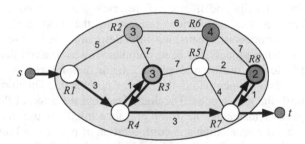

**Figure** Optimal solution with intermediate processing can produce non simple path

Figure depicts a sample network with various processing sites. Each site has an associated cost shown as the number in the node that needs to be taken into account if processing on that site occurs. In the example we are looking for an end to end path that includes two intermediate computations with the constraints that the first computation should be placed on one of the light grey nodes R or R and the second on one of the darker nodes R or R. An optimal solution for this constraint based routing problem taking into account *both* link and processing costs can produce a *non simple path* also known as a *walk* which is a sequence of consecutive edges where a given vertex is being visited more than once. Since such solutions are now possible when considering active processing the signaling protocol must also support such paths. In the following we assume that such explicitly routed paths can be computed according to [ ] and focus on the signaling protocol required for configuring such combined routing and processing paths.

## Explicit Path Establishment

Our proposed *Path Based Routing* (PBR) protocol supports per flow based explicit path establishment for one way unicast flows routed through a predefined list of hops and the installation and configuration of plugin modules along such paths.

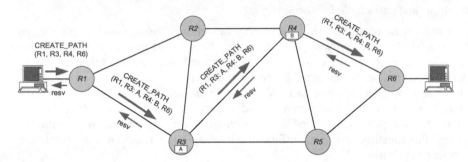

**Figure** Explicit path setup using PBR protocol

As illustrated in Figure the path establishment is based on a *two phase* scheme. In the first phase the protocol verifies whether sufficient resources are available along the *downstream path*. Beginning at the source each node checks whether the required resources are available locally and if granted reserves but does not allocate yet resources and forwards the reservation request to the next node along the downstream path. This process is repeated until the destination node is reached. Once the first phase of the setup process has been completed it is assured that sufficient resources are available. In the second phase the actual allocation of network resources takes place. This happens along the *reverse path* that is on all routers from the destination towards the source node. This includes the installation of flow specific filters such that packets matching the filter are forwarded on the corresponding outgoing interface and the installation and configuration of plugins and binding them to the filter. Once all state has been established along the path the application is informed and can transmit traffic.

If during the first phase a request is refused due to limited resources the path setup process cannot continue and is aborted. The node then sends a reservation release message along the reverse path so that nodes that have already reserved resources can free them. Finally the application is notified that the path could not be established.

The PBR protocol uses TCP as the transport mechanism between PBR peers hop by hop to send control messages for path establishment plugin deployment and release of resources. This guarantees reliable distribution of control messages. PBR uses *soft state* for both path and plugin information being stored on nodes to take into account that network nodes and links are inherently unreliable and can fail. An application that sets up a path is required to refresh the path by sending the setup request periodically otherwise nodes will purge path and plugin state once the time out expires Path tear down works analogous to the path setup process with a release request used instead

### Plugin Deployment

In addition of setting up flow specific routes the PBR protocol allows to install and configure plugin modules on selected nodes To support this feature the path establishment message includes a list of nodes where plugins need to be installed If the requested target address for a plugin matches a node's own address the node first checks whether the referred plugin class has already been loaded into the kernel If it is not present the plugin loader retrieves it from a remote code server and verifies the consistency by checking the module's digital signature [ ] Then the module is loaded into the kernel and linked against the current kernel image Subsequently the PBR daemon creates a new instance of the plugin invokes the configuration method and binds the plugin instance with the filter describing the flow Once the path has been established and the required plugins deployed the application can begin transmitting data which will be forwarded along the path and processed by intermediate plugins

### Message Details

A *CREATE_PATH* message is transmitted by the path initiating router toward the destination to establish an explicitly routed path The message contains the following different subobjects

- *FLOWSPEC* object

  The flow specification describes the format of packets that follow the explicit path described using the tuple <source addr mask dest addr mask source port dest port protocol> Any field can be wildcarded network addresses can be partially wildcarded with a prefix mask

- *EXPLICIT_ROUTE* object

  An explicit route object is encoded as a series of nodes to be traversed In the current implementation the hops must form a strict explicit route

- *PLUGIN* object

  The plugin object describes one or multiple plugins that need to be deployed on a node It contains the address of the target node followed by the plugin name and an initial configuration parameters

The *RELEASE_PATH* message removes a previously established path The PBR protocol also supports the *STATUS* message for the retrieval of path state from remote nodes

### Forwarding Mechanisms for Explicit Path

In the following we describe how we can override Linux's conventional destination based forwarding and perform our own explicit path routing that is move packets along a predefined set of nodes For each flow that requires flow specific forwarding we add a filter entry into netfilter Incoming packets are classified by the netfilter framework and if a flow specific filter matches they are marked with a *next hop neighbor tag*. For each adjacent neighbor there is a special routing table containing a single default route entry pointing to the corresponding neighbor. Packets that have been marked with the next hop tag then use the corresponding routing table and are sent to the appropriate neighbor. To establish a complete path filter entries are added to all nodes along the path.

As discussed above when considering processing sites the path from the source to the destination does *not need to be a straight IP path* anymore where all of the nodes are distinct and no duplicated nodes exist. To support such paths the forwarding mechanism must consider the *incoming port* from where the packet has been received. For that reason flow filters consist of a six tuple including the *incoming interface* for the forwarding decision as well.

Note however that the current implementation is restricted to paths that enter a node via the same interface only once since a node cannot distinguish if a packet has previously traversed the node. This limitation could be overturned if incoming packets would be marked with a tag to be used for subsequent forwarding decisions.

### Example Use of Path Based Routing

In the scenario as illustrated Figure  we want to install an encryption and decryption plugin on selected nodes. When considering processing costs the optimal solution for such a path is *non simple*. The following command establishes a path which routes all traffic matching the filter along the path.

```
pbr create R R R R R R R dport plugin R ENCRYPT init
plugin R DECRYPT init
```

## Related Work

In this section we look at related work both at active network architectures and mechanisms for explicit routing and remote code deployment.

### Programmable Network Architectures

In the context of programmable networks several node architectures have been proposed allowing to dynamically extend the networking subsystem of a router with additional functionality.

The *Active Network Node* [ ] is a NetBSD based architecture that allows code modules called router plugins to be dynamically downloaded and installed into the OS kernel and binding plugin instances to flows. PromethOS's plugin concept has been inspired mostly from this project. While ANN provides many of the concepts implemented PromethOS requires less modification of the original network stack.

*Scout* [ ] proposes a path based approach where the functionality of a standard IP compliant router is decomposed into a sequence of a interconnected components forming a path. Recently Scout has been ported to Linux [ ] however requiring to replace most of the Linux network stack with the Scout implementation.

*NetGraph* [ ] is a the network stack filtering mechanisms for FreeBSD similar to netfilter. The concept is based on hooks offering bidirectional communication for components attached to the network stack. These hooks are freely interconnectable to form a network graph.

*Click* [ ] is an architecture for assembling a router from packet processing modules called elements. Individual elements implement router functions like packet classification, queuing, scheduling, and interfacing with network devices. A router configuration is a directed graph describing the required components and how packets flow through the router. Click configurations are later compiled to machine code providing very good performance. Once defined configurations are static and cannot be tailored at run time, unlike PromethOS plugins. The static approach of Click is overcome by the *Open Kernel Environment Corral* [ ] which makes use of the type safe C programming language *Cyclone* [ ] that is extended by authentication mechanisms for accessing resources.

### Explicit Path Routing and Service Deployment

Several resource allocation protocols capable of supporting applications that request per flow routing and allow functions to be deployed in the network core have been developed. This section briefly describes a few of these protocols.

**Table** Comparison of signaling protocols supporting explicit paths

| | Source Routing | ATM PNNI | MPLS | Beagle | PBR |
|---|---|---|---|---|---|
| Plugin deployment | no | no | no | yes | yes |
| Explicit routing | strict or loose | strict or loose[a] | strict or loose | strict or loose | Strict[b] |
| Looping paths | No[c] | no | no | no | yes |
| Router state | None[d] | VCI VPI entry | MPLS tag entry | RSVP filter entry | netfilter filter |

a PNNI supports hierarchical routing where the source can address the logical group leader representing an aggregation of nodes.
b PBR currently supports only strict routes but loose routes could be easily implemented.
c May be possible but not intended by IP protocol.
d Hop addresses are stored directly in the IP option header. However, due to the option header length limit, the number of hops is restricted to eight.

The *IP source routing option* [ ] provides a means for the source of an IP datagram to supply routing information to be used by intermediate routers. The route data are composed of a series of Internet addresses present in the IP option header. Since there is an upper limit of the option header length, only hosts can be explicitly routed.

The *Private Network to Network Interface* [ ] is a signaling and routing protocol between ATM switches with the purpose of setting up Virtual Connections (VCs). PNNI determines an optimal path satisfying QoS constraints and reroutes connections (crankback) when VC establishment fails. PNNI performs *explicit source routing* in which the ingress switch determines the *entire path* to the destination. Setting up explicit paths is seen as an attractive feature of ATM since each application can have its own specific path requirements. Nevertheless, ATM does not support the concept of processing resources as introduced by active networks.

The *Multiprotocol Label Switching* [ ] approach is based on a label swapping paradigm implemented in the networking layer. MPLS defines two label distribution protocols that support *explicitly routed* paths. CR-LDP [ ] which is an extension of LDP [ ] is peer-to-peer protocol where messages are reliably delivered using TCP and state information associated with explicitly routed LSPs does not require periodic refresh. An explicit route can be *strict* where all nodes are explicitly listed, or *loose* allowing to define paths through portions of the network with partial knowledge of the topology. RSVP-TE [ ] extends the original RSVP [ ] protocol by setting up explicit label switched paths and to allocate network resources (e.g. bandwidth). The *explicit route object* encapsulated in a Path message includes a concatenation of hops describing a strict or loose route. RSVP-TE is based on soft state where the state of each LSP must periodically be refreshed typically every   seconds. CR-LDP and RSVP-TE are signaling protocols that perform similar functions but currently no consensus exists on which protocol is technically superior.

*Beagle* [ ] is a signaling protocol for the setup of structured multi-party/multi-flow applications described by an application mesh. The mesh formulates the resources to be allocated as a network graph. The Beagle protocol is based on RSVP and introduces a new *route constraint object* carrying explicit routing information. In contrast to signaling protocols like MPLS and PNNI Beagle allows applications to allocate computation and storage resources required for delegates, which are application specific code segments that execute on routers.

The PBR protocol has specifically been designed for active networks. It allows the deployment of new code on routers and the setup of explicitly routed paths supporting also looping paths such that the same processing site can be visited multiple times.

## Conclusions

In this paper we have presented PromethOS, an extensible and modular architecture for integrated services routers. PromethOS allows to dynamically load plugins at runtime into the kernel to create instances of plugins and to bind plugin instances to individual flows. The path-based routing protocol establishes explicitly routed paths and installs plugins on selected nodes. We freely distribute our source code with the intent of providing the research community with a services platform to build upon.

Currently, PromethOS is being extended to provide resource control mechanisms for plugins in kernel space. We focus on aspects of memory consumption, processor cycles and bandwidth on both general purpose and network processors.

# References

[ ] ATM Forum Technical Committee, "Private Network Network Interface Specification Version" "March"

[ ] Awduche D, Berger L, Gan D, Li T, Swallow G and V Srinivasan, "RSVP TE Extensions to RSVP for LSP Tunnels" *RFC* December

[ ] Andersson L, Doolan P, Feldman N, Fredette A, Thomas B, "LDP Specification" *RFC* January

[ ] Andy Bavier, Thiemo Voigt, Mike Wawrzoniak, Larry Peterson, Per Gunningberg, "SILK: Scout Paths in the Linux Kernel" Department of Information Technology, Uppsala University

[ ] Herbert Bos and Bart Samwel, "The OKE Corral: Code Organisation and Reconfiguration at Runtime using Active Linking" *IWAN* December

[ ] Braden R, Zhang L, Berson S, Herzog S and S Jamin, "Resource ReSerVation Protocol RSVP – Version Functional Specification" *RFC* September

[ ] K L Calvert et al, "Architectural Framework for Active Networks Version" *DARPA Active Network Working Group Draft* July

[ ] Prashant Chandra, Allan Fisher, Peter Steenkiste, "Beagle: A Resource Allocation Protocol for Advanced Services Internet" Technical Report *CMU CS* August

[ ] Sumi Choi, "Plugin Management" Washington University in St Louis *Technical Report WUCS*

[  ] Sumi Choi, Jonathan Turner, Tilman Wolf, "Configuring Sessions in Programmable Networks" In Proceedings of Infocom March

[  ] A Cobbs, "All About NetGraph" http://www.daemonnews.org netgraph.html

[  ] Cyclone AT T Research Labs and Cornell University http://www.research.att.com projects cyclone

[  ] Decasper D, Dittia Z, Parulkar G, Plattner B, "Router Plugins – A Modular and Extensible Software Framework for Modern High Performance Integrated Services Routers" *Proceedings of ACM SIGCOMM* September

[  ] Jamoussi B et al, "Constraint Based LSP Setup using LDP" *RFC* January

[  ] Ralph Keller, Sumi Choi, Dan Decasper, Marcel Dasen, George Fankhauser and Bernhard Plattner, "An Active Router Architecture for Multicast Video Distribution" *Infocom* Tel Aviv March

[  ] Eddie Kohler, Robert Morris, Benjie Chen, John Jannotti, M Frans Kaashoek, "The Click modular router" *ACM Transactions on Computer Systems* August pages

[  ] David Mosberger, Larry Peterson, "Making Paths Explicit in the Scout Operating System" *Operating Systems Design and Implementation* pages

[  ] J Postel, "Internet Protocol" *RFC*

[  ] PromethOS website http://www.promethos.org

[  ] Rosen E, Viswanathan A, Callon R, "Multiprotocol Label Switching Architecture" *RFC* January

[  ] Rusty Russell, "Linux NetFilter Hacking HOWTO" http://www.netfilter.org

# The OKE Corral: Code Organisation and Reconfiguration at Runtime Using Active Linking

Herbert Bos and Bart Samwel

LIACS, Leiden University, Niels Bohrweg 1, 2333 CA, The Netherlands
{herbertb,bsamwel}@liacs.nl

**Abstract.** The OKE Corral is an active network environment which allows third-party active code to configure an active node's code organisation at any level, including the kernel. Using the safety properties of an open kernel environment and a simple 'Click-like' software model, third parties are able to load native code anywhere in the processing hierarchy and connect it to existing components at runtime.

## 1 Introduction

For reasons of safety, most active networks (ANs) tend to sandbox active code in user space, either locally or at a remote node. Moreover, such code is often interpreted, which slows down its performance considerably. Even in non-active environments interpreters are frequently used whenever application-specific code is loaded in the kernel. A well-known example is found in BSD packet filtering.

In previous work, however, we have shown how the open kernel environment (*OKE*) provides a safe, resource-controlled programming environment which allows fully optimised native code to be loaded in a Linux kernel by parties other than `root` in a safe manner [BS02]. In this paper, we describe how the *OKE* was used to develop an environment for building high-speed ANs allowing third parties to load and configure native code anywhere in the processing hierarchy. An implementation of this environment is found in the *Corral* (Code Organisation and Reconfiguration at Runtime using Active Linking). The contribution of this work is that three existing technologies in programmable networks (open kernels, the 'Click router' model, and ANs) are combined to provide a safe platform for fast packet processing in the kernel of a common operating system while explicitly separating control and data. In the *OKE Corral* high-speed packet processing is managed by slow-speed control code. It can be summed up as follows:

1. We borrow the LEGO-like software model advocated by the 'Click router' project in [CM01] to build both fast *data* paths and slower *control* paths.
2. One of the components on the control path is an AN runtime.
3. The configuration/implementation of the paths is controlled by third-party code executing either on the node itself, e.g. in the form of active applications (AAs), or at a remote site.
4. The *OKE* ensures that new kernel-level path components are safe.

J. Sterbenz et al. (Eds.): IWAN 2002, LNCS 2546, pp. 32–47, 2002.

Although the individual components may not be new, to the best of our knowledge there does not exist any system that provides the following combination of features in a commonly used operating system: (a) programmability of both kernel and user space with fully optimised native code, (b) while still providing full resource control and safety with respect to memory, CPU, available API, etc., and (c) allowing for flexibility in the amount of programmability permitted on a node, and (d) where control over fast native code components is exercised by slow-speed active applications (AAs), (e) by means of a simple 'Click-like' programming model,

Not all issues concerning the use of the *OKE* for ANs are addressed in this paper. In particular, node heterogeneity and scaling of trust relationships to large networks are not addressed. However, while the *OKE* relies on trusted compilers, the issue of trusting compilers in a remote domain is non-trivial and important. We will briefly discuss this in Section 3.3.

The remainder of this paper is organised as follows. The *OKE Corral* architecture is explained in Section 2, the prototype implementation of the architecture is discussed in Section 3. This is evaluated in Section 4. Related work can be found in Section 5 and conclusions are drawn in Section 6.

## 2 Architecture

As illustrated in Figure 1, the *OKE Corral* builds on three technologies: (1) the *OKE*, (2) one or more AN runtimes, and (3) packet channels that implement control and data paths. To the right of the architecture we have indicated the approximate mapping of *OKE Corral* components on the DARPA reference architecture of an active node. By necessity this is only an approximation. For instance, depending on the configuration the kernel may or may not be dynamically programmed (i.e. run AAs in its execution environment).

### 2.1 Corral Terminology

The terminology in the *OKE Corral* roughly follows that of the Click-router project, although there are some differences. What are called processing elements in Click are called 'engines' in the *Corral*. They may have multiple input and output ports that can be logically attached to other elements to form connections (drawn as arrows). A data transfer whereby the source element takes the initiative is called a *push* operation, while a transfer initiated by the destination is called a *pull*. Connections are either of the *push* or the *pull* type. Engines are normally also either 'push' or 'pull', but a hybrid form, known as *pull2push* is also possible. A pull2push engine pulls data from a source and pushes it to a destination, changing the pull into a push.

The inverse of a pull2push element is a queue, as it accepts pushes on its input, and pulls on its output, and thus may be termed a *push2pull* element. Queues may be filled and emptied by more than one engine. As shown in Figure 1, engines and queues are connected and disconnected via control operations. The

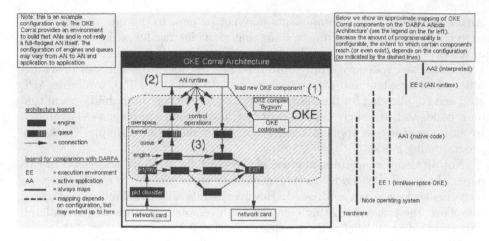

**Fig. 1.** Overview of the *OKE Corral*

path followed by a particular packet is known as a 'channel'. In Figure 1, the entry engine, the two boxes to the right of it, and the exit engine form a channel.

In contrast to the Click approach, queues and engines may reside in the kernel, in user space, or even on remote machines. Wherever they reside is known as the queue or engine's "domain". Similarly, they may exist either inside the *OKE* (in which case they are subject to checks and resource limits), or as native, unprotected code. The packet classifier in the figure determines which packets are relegated to the AN's channels. It is really part of the *OKE* environment setup code (ESC) for the AN, but it lies beyond the reach of the AN and because of this it is drawn outside of the *OKE* box.

## 2.2   OKE, AN Runtime, and Channel Interaction

When an AN runtime is instantiated, it is initially provided with a channel consisting of two engines: the entry engine and the exit engine. All the AN's packets are first pushed to the entry engine, which automatically leads to a push to the exit engine. Each of the pushes is implemented as a function call, executed immediately and in the same thread of control.

The AN is allowed to disconnect the two engines and reconnect them (at runtime), e.g. to new components inserted between them. For example, to receive all packets in its runtime, a trivial AN implementation might: (1) disconnect the two engines, (2) reconnect the entry engine to a queue, (3) implement an engine's interface for the runtime (essentially making the runtime an engine which 'pulls' packets from the queue and pushes them up into the runtime), and (4) implement the runtime's **send** operation as a push to the exit engine. All incoming packets classified as AN traffic are now automatically pushed onto the queue, and from there pulled up into userspace.

The AN is given a set of standard components (engines and queues) with which to build channels (subject to the privileges given to the AN). These standard components can be highly optimised so as to incur few checks at runtime. In addition, the AN is able to load entirely *new* components. In case native code is to be loaded in the kernel, the *OKE* is used to ensure safety. As discussed in Section 3.1, the *OKE* is able to restrict code according to the loading party's credentials. In *OKE* terminology, the credentials presented by a client's are defined as that client's *role*. Thus, a party in a highly untrusted role may be allowed to load code with very few privileges and many dynamic checks, while a party in a highly trusted role may benefit from a more relaxed security policy.

## 2.3 Control and Data Channels

Using the above techniques, an AN is able to build fast channels where processing is done in optimised native code and where the next processing stage is always just a function call away. At the same time we also use channels to implement slow-speed control paths which commonly lead to AN runtimes in user space (or even remote hosts) and which are used to carry the active packets containing the control code. Given the appropriate privileges they are able to replumb, or add new elements to, the data-path at runtime. Thus, the amount of data-path programmability allowed is configurable, which is useful if active networks are to scale.

# 3  Implementation

## 3.1  The Open Kernel Environment

Although the running of third-party code in the kernel normally violates security constraints, it would be useful from a performance perspective. In the *OKE*, instead of asking whether or not a party may load code in the kernel, we ask: what is such code allowed to do there? *Trust management* determines the privileges of user and code, both at compile time and at load time. Based on these privileges a *trusted compiler* enforces extra constraints on the code. In the following we briefly describe the *OKE*'s two main components: the code loader (CL), and the *bygwyn* compiler (as previously presented in [BS02]). A pre-release of the *OKE* as well as an extended version of this paper can be downloaded from www.liacs.nl/~herbertb/projects/oke/.

The CL accepts object code, together with authentication and credentials, from any party. It checks the credentials against the code and the security policy and loads the code if they match (Figure 3). Trust management is based on KeyNote [BFIK99]. At start-up, the CL loads a security policy, containing the keys of any party permitted to load certain types of modules. These parties are then able to delegate trust to other clients by way of credentials containing containing the 'rights' that are granted, e.g. the right to 'load modules of *type* X or Y, but only under *condition* Z'. A 'type' here denotes the set of privileges given to the code, e.g., the interface to the kernel, the amount of CPU time, etc. The 'condition' contains environment-specific stipulations, e.g., 'only valid during office hours'. A module type is instantiated when source code corresponding to it is compiled. The trusted compiler generates an unforgeable 'compilation record' which proves that module $M$ was compiled as type $T$ by this compiler.

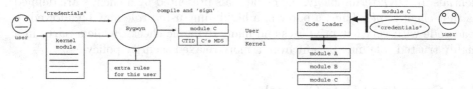

**Fig. 2.** User compiles kernel module   **Fig. 3.** User loads module in the kernel

It is crucial that we guard against malicious or buggy code in the kernel. What we have tried to avoid, however, is the definition of yet another safe language which is only useful for implementing filters, say and/or runs inside an interpreter, as such a language necessarily restricts towards the lowest common denominator. Instead we would like to have a single language that is automatically restricted on the basis of explicit privileges. A single language is preferable to many special-purpose languages for many reasons, e.g., consistency, learnability, maintainability, flexibility, etc. Moreover, using a language like C would facilitate the interfacing of third party code to the rest of the kernel.

We therefore allowed a C-like language to be restricted in such a way that, depending on the client's privileges more or less access is given to resources, APIs and data (and/or more or less runtime overhead is incurred). As C itself is not safe and the possibilities of corrupting a kernel using C are endless, we opted for *Cyclone*, a crash-free language derived from C which ensures safe use of pointers and arrays, offers fast, region-based memory protection, and inserts few runtime checks [JMG⁺02]. However, for true safety and speed, using Cyclone was not sufficient. For example, we had to add an entirely new garbage collector to deal with pointers to kernel memory. Other hard problems (e.g., resource limitation, module termination, and memory/pointers sharing) are also not solved by Cyclone. We therefore created our own dialect which we call 'OKE-Cyclone'.

The restrictions are enforced by a trusted compiler, known as *bygwyn*, (named after a track by the Rolling Stones: 'You can't always get what you want, but you get what you need'). *Bygwyn* is customisable, so that in addition to its normal language rules, it is able to apply extra rules as well. For example, we allow one to remove constructs from the language. If after such a restriction the compiler encounters the forbidden construct, it generates an error.

The key idea is that the customisations for a user's program depend on the user's role: users present credentials to the compiler, and these credentials determine which rules are applied (Figure 2). Customisation types have unique identifiers, called customisation type identifiers (CTIDs). After compilation, *bygwyn* generates a signed compilation record containing both the CTID and the MD5 of the object code, explicitly binding the code to a type. Given this, we allow security policies to be specified of the form 'a user with authorisation X is allowed to load code that is compiled with customisation Y'. Once loaded, the code runs natively at full speed.

Depending on the users' roles, they get access to the rest of the kernel via an API containing the routines which they may call (e.g., students in a course on kernel programming may get access to different functions than third-party network monitors). The routines are linked with the user code and reflect its role. In other words, the API is used to *encapsulate* the rest of the kernel (Figure 4). In the figure, some function calls are relegated to a wrapper, while others may be called directly.

We now briefly mention some of the mechanisms we implemented for making the OKE-Cyclone dialect safe for use in the kernel.

1. We perform global code analysis to decrease the number of dynamic checks.
2. Environment setup code (ESC) containing the customisations is automatically prepended. It declares kernel APIs and other functions and variables and leaves the untrusted code with only the safe API (wrappers mostly). It also provides wrapper code for resource cleanup and safe exception catching. The ESC can configure this wrapping using a new `wrap extern` construct: *bygwyn* detects all potential entry points to the untrusted code and automatically wraps them using code declared by the ESC.
3. Certain language constructs can be removed from the programmer's repertoire using a new `forbid` construct (examples include: `forbid extern "C"`, `forbid namespace`, and `forbid catch`).
4. A unique, randomly generated namespace is opened to prevent namespace clashes and unauthorised imports of symbols from other namespaces.
5. The stack usage of the code can be restricted to a limit defined in the ESC.
6. CPU usage is limited by using a modified timer interrupt. When a module has not finished on time, an exception is thrown and the module is removed. Code misbehaving in other ways is likewise removed.
7. Cyclone's region-based memory protection mechanism was extended with a new region `'kernel`, to distinguish between kernel-owned and module-owned memory regions and a new garbage collector was implemented to ensure that pointers from the *OKE* modules to kernel memory (which may be manipulated by kernel functions) are memory safe, and freeing of module memory is handled correctly.
8. Specific fields of kernel structures shared with untrusted code can be statically protected by making making them `locked`. A `locked` member cannot be used in calculations, it cannot be cast to another type, no other type can be cast to it, no pointer dereferences can take place, and no structure members can be read. Basically, its is limited to copying, and it cannot be read. This technique reduces the need to anonymise data at run-time.

## 3.2   Channels

The concept of clicking kernel components together to create new functionality is a tried and useful practice. The *x-Kernel*, first proposed in the late 80s, provided mechanisms to statically stack network protocols in this way [HP91]. Similarly, the *STREAMS* abstraction, proposed even earlier, allowed protocol stacks to be composed dynamically [Rit84]. This work was influenced by all such approaches and in particular, as mentioned earlier, by the Click software router. The *OKE* channel elements all have simple interfaces that are implemented in either C or OKE-Cyclone. Each channel element carries pointers to its own state, as well as to both blocking and nonblocking implementations of the pull and push operations. In this section we describe the main features of engines and queues. In essence, queues and engines have unique identifiers and communicate by pulling and pushing data from and to each other's ports. A push or pull connection is typed, so that only specific items may be pushed or pulled on a connection. The types range from simple types such as integers and octets to composite types (e.g., IP packets). We have not addressed the issue of how specific engines or queues to connect to are discovered or located. This should not be a problem for a handful of elements that we loaded ourselves on a single node, but for large-scale deployment such functionality would be very useful.

Queues in the default implementation are strictly FIFO (producer/consumer on a circular buffer). More complex queueing schemes can be constructed using multiple FIFOs, or by providing an implementation of custom push and pull functions. Queues are passive elements. They respond to **push** and **pull** operations, but never initiate actions themselves. In contrast, engines are active elements. Apart from push and pull, they also provide a control interface (Figure 1), and a **run** method which is called when it is scheduled. The control interface contains the **connect** methods needed to link engines to other engines or to queues. These methods take as arguments (among other things) the unique name of the target element, as well as the port and the port direction (input or output). Queues do not provide such methods: they are managed by engines.

Engines and queues can be (dis-)connected at *runtime*. As such, the connections between them are not built into their logic. Instead, the control API allows explicit replumbing of the components. As it is dangerous to replumb an element when it is active (e.g., about to push a packet to an engine we would like to disconnect), these activities are protected by a 'readers-writers' solution: many different data-path actions may be taking place at any time, but management operations such as **connect** require exclusive access.

Engines and queues are tied to a *domain*. Currently, possible domains are: *userspace*, *kernel*, and *remote*. Elements in the same domain communicate by pushing or pulling simple types directly, or complex types by passing pointers, making communication within the same domain quite efficient. It is also possible to place engines and queues in different domains. For this purpose we use simple marshalling techniques commonly used in remote procedure calls. For example, if an engine in domain $D_1$ wants to push a packet to queue $Q$ in domain $D_2$, it really calls the **push** operation on a local proxy $Q_{proxy}$ (also known as 'stub'). $Q_{proxy}$ is initialised with a set of routines that enables it to connect to the remote implementation of $Q$. It marshalls the packet and initiates a 'remote' procedure call to push the packet on the 'remote' queue. 'Remote' here means a different domain, which could easily reside on the same host. Default proxies and marshalling routines have been written which are expected to suffice for most applications. Even so, the scheme can be easily extended.

Packet traversal in the *Corral* is as follows. Once a packet is classified (by the classifier in Figure 1) as belonging to the AN, it is pushed on the AN's entry engine and follows the data-path determined by the AN's engines and queues. Some of the fields in the packet may be protected against access violations using the `locked` keyword. Locked fields cannot be pushed across domains. The entry engine pushes the packet to the next engine and so on, until one of the following three events has occurred: (1) the packet is dropped, (2) the exit engine is reached and the packet has been sent, or (3) an intermediate queue has been reached.

### 3.3    The Active Network

The AN runtime is derived from a home-grown active network, which is capable of running either a Java or a Tcl execution environment. For the *OKE Corral* implementation we have limited ourselves to the Tcl implementation. The runtime provides a simple environment for AN experiments and permits code loading both in-band and out-of-band. It consists of an interpreter and a fairly extensive set of operations specific to the AN. This is called the *core set*, which is implemented in C. The core set contains elementary operations, e.g. functions to access received packets and to find the load on specific links, etc. It also contains a `send` operation for transmitting a packet. Packets are stored in packet buffers, of which there is a fixed number. One of the buffers is designated the 'current' buffer and this is used to receive the next packet. A number of operations in the core set is responsible for managing the buffers, e.g. to set the current buffer, to execute safe `memcpy` and `memmove` operations, etc. An additional library that is fully implemented in Tcl contains a large number of functions that are commonly used, as well as wrappers around the core set.

The runtime back-end was modified to sit on top of the *OKE* channels. More correctly, by implementing the engine interface, the runtime really becomes an engine $E_R$ itself. $E_R$ initialisation code disconnects the packet entry and exit engines assigned to it and reconnects the entry engine to a kernel-domain queue. It also connects $E_R$ to the other end of the queue for inbound traffic and to the packet exit engine for outbound traffic.

After initialisation, the active code in the runtime is responsible for the management and control of the engines and queues in its channels. For example, operations were added to the AN's repertoire to to connect or disconnect all elements under its control. Depending on the AN, bootstrap kernel modules containing pre-installed engines and queues may be loaded at initialisation. The components in such modules can be used by the AN to construct new data-paths. They may be highly efficient, e.g. written in C and containing few runtime checks.

There are also commands to enable the active code to add entirely new components (engines and queues) to the data-path. In the following discussion we assume that the target domain for the new components is the kernel, since this presents the most severe security risks. For the purpose of loading data-path components, the active code refers to new kernel modules on a remote webserver. Similarly, it uses URLs to refer to the credentials. Next, the module and the credentials are both loaded and offered to the *OKE* codeloader. Provided the credentials match, the module is pushed into the kernel. At that point the AN is able to manage the new engines and queues in exactly the same way as the pre-installed components.

When loading new components in the data-path, safety is guaranteed by the *OKE*. This means not only that the code *must* be written in OKE-Cyclone, but also that the compiler that compiles it must be *trusted*. We have not addressed the issue of whether under what circumstances compilers in remote domains can be trusted. We call this the 'trust propagation' problem. One possibility is to have a well-known group of trusted compilers that are accepted by many sites (the "VeriSign model"). Alternatively, we might store the code in source format and have a *local* (and presumably trusted) compiler generate the object code anew just prior to loading it. We are currently exploring and evaluating these and other solutions.

Another issue concerns the authorisation of requests to load code in the kernel. Normally, when a client tries to load a module in the kernel it is required to authenticate every such request by signing a number of items with its private key. However, if code is running on an unknown active node, clients may be reluctant to send their private keys there for obvious reasons. In the *OKE Corral* model, we have 'single sign-on' behaviour. In other words, the identity of the client is established when the active code is loaded on the runtime. From that point onwards, this is the identity that is used in `load`, `connect` and other requests. The active code is not required to sign anything.

## 4   Results

We do not think that the number of packets per second that can be handled is a relevant measure in evaluating the *OKE Corral*, for two reasons. First, such numbers often say more about the traffic *capture* (e.g. polling or interrupt-driven) than about the *processing* [ST93]. Second, we are really interested in how the *Corral* compares to typical ANs and this concerns primarily the nature of code execution: in-kernel native code versus interpreted code in userspace. For the number of packets per second that *can* be processed with a channel-based system, please refer to [CM01].

Instead, we measure the performance of the data-path components and compare the results with alternative implementations. All measurements were taken on a PIII 1GHz PC running a Linux-2.4.18 kernel. The overhead of a push from entry engine to exit engine without any processing takes roughly 250 nsecs (including all locks and sanity checks). The applications used for the comparison are in the domains of transcoding (application $T$) and monitoring (application $M$). Both are considered components on the data-path. In the *OKE Corral* version of the experiment, they are implemented in OKE-Cyclone, and loaded in the kernel by the active control code in userspace. $M$ implements a packet sampler which is meant to push 1% of all packets on a queue which is read by the AN monitoring application in userspace using a pull. On request (pull), $M$ also reports the total number of bytes of all packets that passed through $M$ since the last report. $T$ resamples audio packets to a lower quality (containing half the bits) and thus works on the entire payload. For this reason, $T$ also requires a recalculation of the IP and UDP checksums. Both types of applications may operate on the same packets. In fact, there are 4 types of packet, all of which are UDP with destination ports $p_0$, $p_1$, $p_2$, and $p_3$. The experiment is illustrated in the leftmost illustration of Figure 5. Packets for port $p_0$ are subject to both transcoding and monitoring. Packets for $p_1$ are subject to transcoding but not to monitoring, i.e. they are pushed directly to the exit engine by the transcoder engine. Destination $p_2$ packets will pass *through* the transcoder, but are not touched by it. Instead they are moved straight to the monitoring engine. Packets for $p_3$ are neither resampled nor monitored, but do pass through the entry engine, the transcoder and the exit engine.

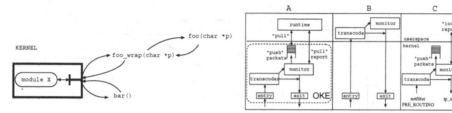

**Fig. 4.** Kernel encapsulation     **Fig. 5.** The three scenarios used

We evaluate 3 different implementations: (A) all components in the *OKE*, (B) all components in the AN runtime, and (C) all components in in-kernel C, as shown in Figure 5. All three versions are possible in the *OKE Corral*, but we are most interested in solution (A), as it provides maximum flexibility while still running natively in the kernel. We measure time between packet entry at the Linux netfilter hook to the time that we send the packet (or queue it for userspace).

The results are shown in Figures 6-8. As expected, we see in all figures that, since the entire payload must be processed for $T$, the overheads for $p_0$ and $p_1$ packets strongly depend on the packet sizes. The $p_2$ and $p_3$ graphs on the other hand are basically flat, as we do not even need to recalculate the checksums for these packets. We also observer that in the Tcl implementation the effect of monitoring is no longer visible. This is due to the enormous overhead introduced by the interpreter and context switching.

No manual optimisation was used in any of the implementations. Moreover, there exist much faster AN runtimes than the one we have used. However, in previous work we measured that a copy from kernel to userspace using an `ioctl` channel takes roughly 2 $\mu s$, and considerably longer with `libipq` (8 $\mu s$ on average). If a copy to userspace is needed, it will be difficult to optimise away this overhead. A copy back to the kernel takes approximately the same amount of time, so regardless of the speed of the C code, we lose $4\mu s$, just on the copies.

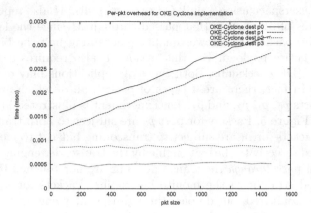

**Fig. 6.**  Scenario A: in-kernel in-Cyclone

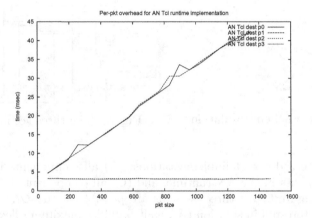

**Fig. 7.**  Scenario B: in-userspace (AN)

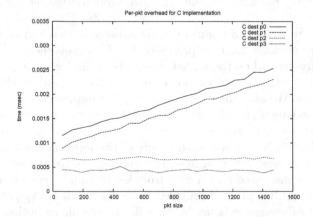

**Fig. 8.**  Scenario C: in-kernel in-C

This overhead alone exceeds the total time needed by the OKE-Cyclone implementation.

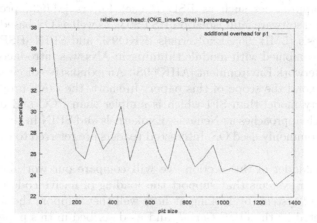

**Fig. 9.** Overhead of processing packet $p_1$ in the OKE compared with C

In Figure 9 we also plot the relative overhead of performing the transcoding application in the *OKE* instead of native C. Concretely, the figure plots the ratio computed by $(\frac{T_{Cyclone}}{T_C} * 100 - 100)$ for the $p_1$ packet times shown in Figures 6 and 8. It is interesting to note that the overhead per byte decreases as the packet size increases. This is caused by the fact that the fairly substantial one-time overhead is ammortised over a large number of bytes. The overhead of the implementation with the AN in the datapath is orders of magnitude and therefore not plotted.

For now, we conclude that the difference in performance between the AN implementation and either of the other two implementations is orders of magnitude. Between the *OKE* and the 'pure C' implementation the difference is roughly 25%. A substantial gain in performance can be achieved by employing the *OKE* in ANs. However, even if the speed of pure C is required, active code is still able to control and manage these components, and to build new applications by clicking together elements from a predefined set.

## 5   Related Work

Organising AN software in a hierarchical fashion is advocated in many active network projects, e.g. SwitchWare [AHK+98]. Such approaches differ from the *OKE Corral* in that they are mostly concerned with (interpreted) user space code for all loadable extensions. Clicking components together to form channels is equally common in ANs. A good example is CANEs, which allows extensions to be injected in predefined locations on the data-path [MBC+99]. A third aspect, the separation of control and data path in programmable networks has also been advocated in a number other projects, e.g. SwitchWare at UPenn and the work on programmable network control in Cambridge [BIML01].

Many projects target safety in operating systems (OSs). These include language-based approaches such as BSD Packet Filters [MJ93], proof carrying code[NL96] and software fault isolation[RSTS93], as well as OS-based approaches such as Nemesis [LMB+96], ExoKernels [EKO94], and SPIN [BSP+95]. Trust management combined with module thinning in ANs was introduced in the Secure Active Network Environment [AHK+98]. An exhaustive discussion of these projects is beyond the scope of this paper. In short, the *OKE* provides a more complete safety model than SFI which is simpler than PCC and distinguishes itself from such approaches as Nemesis, Exokernels and SPIN in that it is implemented on a commonly used OS. Interested readers are referred to the discussion in [BS02].

In the remainder of this section, we will compare our work briefly with a number of other systems that support the loading of native code in the kernel of an operating system, by looking at how well they support the following ten features targeted by the *OKE Corral* (and as described in this paper):

1. The system explicitly supports 3rd party code in the kernel.
2. The kernel is fully programmable, although if needed, we are able to restrict access to specific APIs, data, etc., at compile time.
3. Resource control is enforced for CPU, memory, etc.
4. Safety is enforced in the sense that a module is not able to crash, dereference NULL pointers, inadvertently free kernel memory it points to, etc.
5. Data channels are composed of LEGO-like components (like in Click).
6. Configuration of these channels is possible at runtime.
7. Data and control are explicitly separated.
8. AAs in the form of capsules are able to configure the data channels to the point of loading and connecting new native code components.
9. Out-of-band loading of AAs in the kernel is supported.
10. The system is implemented on a common OS.

Note that we do not aim for a true comparison of these very different systems. We only look at how well other approaches support some of the more attractive features of the *OKE Corral*. The results of the comparison are shown in Table 1. Below we discuss the projects mentioned in the table.

We have been strongly influenced by the Click router's LEGO-like organisation of forwarding code [CM01]. Although we didn't use the Click code directly, we implemented a very similar system (in C). However, whereas Click components are assumed to reside in the same domain (e.g. the kernel), we permit them to be distributed at will over kernel, user-space and even remote machines.

Our processing hierarchy resembles that of the 'extensible router' [NLA+02]. In particular, SILK also provides fast kernel data-paths with support for resource accounting. However, it does not provide safety. The code loading in our work somewhat resembles that of ANN [DPP99]. In ANN active code is replaced by references to modules stored on code servers. On a reference to an unknown code

| | Feature | SILK | ANN | PromethOS | SPIN | FLAME | Click | OKE Corral |
|---|---|---|---|---|---|---|---|---|
| 1 | 3rd party code in krnl | ++ | -- | - | ++ | ++ | -- | ++ |
| 2 | full krnl programmability + restriction possibilities | +/- | +/- | + | ++ | - | +/- | ++ |
| 3 | resource control | + | - | - | + | + | - | ++ |
| 4 | safety guarantees | -- | - | - | ++ | + | -- | ++ |
| 5 | LEGO-like components | + | - | - | - | - | ++ | ++ |
| 6 | dynamic configuration | ++ | + | + | ++ | + | - | ++ |
| 7 | separation control/data | ++ | - | ++ | 0 | + | + | ++ |
| 8 | AA: capsules load in krnl | - | - | - | 0 | 0 | 0 | ++ |
| 9 | AA: out-of-band loading | ++ | + | ++ | 0 | ++ | + | ++ |
| 10 | common OS | ++ | ++ | ++ | -- | ++ | ++ | ++ |

**Table 1.** *OKE Corral* features compared with other systems. Explanation of symbols: '+' = strong support, '-' = weaker, '+/-' = partly, '0' = not applicable to this system.

segment in a node, the native code is downloaded, linked and executed. Similarly, a recent project called PromethOS described elsewhere in these proceedings, supports kernel plugins with explicit signalling for plugin installation [RLAB02]. Neither approach targets safety as aimed for by the *OKE*.

SPIN, which builds on the safety properties of Modula-3, is close in spirit to the work presented here. However, unlike the *OKE*, SPIN does not control the heap used by 'safe' kernel additions. Additionally, it is not a commonly used OS.

Early work on the use of the Cyclone for kernel work and KeyNote for policy control was demonstrated in FLAME [AIM+02] which is similar to the *OKE* and a good example of how similar principles are used for different goals. FLAME is aimed at safe network monitoring and not on fully programmable kernels. In contrast, the *OKE* provides the necessary features for general-purpose kernel extensions, with a focus on customisability. FLAME provides little flexibility in the restrictions placed on a module, and full interaction between the module and the kernel (e.g., using pointers) is not allowed. While essential to the *OKE*, neither of them are needed in FLAME.

# 6   Conclusions

The *OKE Corral* combines a 'Click-like' software model with an open kernel environment under control of an active network while maintaining strict separation of control and data plane. Performance varies with the programmability desired. At one extreme, only the control plane is programmable, while data-paths are composed of highly optimised 'standard' components. At the other extreme, the 'capsule' approach can be supported. In between these two extremes, but closer to the former, we have the *OKE* channels. For flexibility, the different kinds of programmability may be mixed, so that capsules, pre-defined and third-party components all interact to build data and control flows.

# Acknowledgements

We are indebted to Lennert Buytenhek for his help on the thornier issues in kernel hacking and Nadia Shalaby and the reviewers for their excellent feedback.

# References

[AHK+98]  D. Scott Alexander, Michael Hicks, Pkaj Kakkar, Angelos Keromytis, Marianne Shaw, Jonathan Moore, Carl Gunter, Trevor Jim, Scott M. Nettles, and Jonathan Smith. The SwitchWare active network implementation. In *Proceedings of the 1998 ACM SIGPLAN Workshop on ML*, 1998.

[AIM+02]  K. G. Anagnostakis, S. Ioannidis, S. Miltchev, J. Ioannidis, Michael B. Greenwald, and J. M. Smith. Efficient packet monitoring for network management. In *Proc. of NOMS'02*, April 2002.

[BFIK99]  M. Blaze, J. Feigenbaum, J. Ioannidis, and A.D. Keromytis. The KeyNote trust-management system version 2. *NWG RFC 2704*, September 1999.

[BIML01]  Herbert Bos, Rebecca Isaacs, Richard Mortier, and Ian Leslie. Elastic networks: An alternative to active networks. *JCN (Special Issue Programmable Switches and Routers)*, 3(2):153–164, June 2001.

[BS02]  Herbert Bos and Bart Samwel. Safe kernel programming in the OKE. In *Proceedings of OPENARCH'02*, New York, USA, June 2002.

[BSP+95]  B.Bershad, S.Savage, P.Pardyak, E.G.Sirer, D.Becker, M.Fiuczynski, C.Chambers, and S.Eggers. Extensibility, safety and performance in the SPIN operating system. In *Proc of SOSP-15)*, pages 267–284, 1995.

[CM01]  Benjie Chen and Robert Morris. Flexible control of parallelism in a multiprocessor pc router. In *Proc. of USENIX Annual Technical Conference (USENIX '01)*, pages 333–346, Boston, Massachusetts, June 2001.

[DPP99]  D. Decasper, G. Parulkar, and B. Plattner. A scalable, high performance active network node. *IEEE Network*, January 1999.

[EKO94]  Dawson R. Engler, M. Frans Kaashoek, and James W. O'Toole Jr. The exokernel approach to extensibility. In *Proc. of OSDI'94*, page 198, Monterey, California, November 1994.

[HP91]  N. C. Hutchinson and L. L. Peterson. The x-kernel: An architecture for implementing network protocols. *IEEE Trans. on Software Engineering*, 17(1):64–76, 1991.

[JMG+02]  Trevor Jim, Greg Morrisett, Dan Grossman, Michael Hicks, James Cheney, and Yanling Wang. Cyclone: A safe dialect of C. In *Proceedings of USENIX 2002 Annual Technical Conference*, June 2002.

[LMB+96]  Ian Leslie, Derek McAuley, Richard Black, Timothy Roscoe, Paul Barham, David Evers, Robin Fairbairns, and Eoin Hyden. The Design and Implementation of an Operating System to Support Distributed Multimedia Applications. *JSAC*, 14(7), September 1996.

[MBC+99]  S. Merugu, S. Bhattacharjee, Y. Chae, M. Sanders, K. Calvert, and E. Zegura. Bowman and canes: Implementation of an active network, 1999.

[MJ93]  Steven McCanne and Van Jacobson. The BSD Packet Filter: A new architecture for user-level packet capture. In *Proceedings of the 1993 Winter USENIX conference*, San Diego, Ca., January 1993.

[NL96]  George Necula and Peter Lee. Safe kernel extensions without run-time checking. In *Proceedings of OSDI'96*, Seattle, Washington, October 1996.

[NLA+02]  N.Shalaby, L.Peterson, A.Bavier, Y.Gottlieb abd S.Karlin, A.Nakao, X.Qie, T.Spalink, and M.Wawrzoniak. Extensible routers for active networks. In *DARPA AN Conference and Exposition*, June 2002.

[Rit84]  D. M. Ritchie. A stream input-output system. *AT&T Bell Labs Technical Journal*, 63(8):1897–1910, 1984.

[RLAB02]   R.Keller, L.Ruf, A.Guindehi, and B.Plattner. PromethOS: A dynamically
           extensible router architecture for active networks. In *Proc. of IWAN 2002*,
           Zurich, Switzerland, December 2002. Springer.
[RSTS93]   R.Wahbe, S.Lucco, T.E.Anderson, and S.L.Graham.   Efficient software-
           based fault-isolation. In *Proc. of SOSP'93*, pages 203–216, December 1993.
[ST93]     Jonathan M. Smith and C. Brendan S. Traw.  Giving applications access
           to Gb/s networking. *IEEE Network*, 7(4):44–52, 1993.

# Lightweight Thread Tunnelling in Network Applications

Austin Donnelly

University of Cambridge, Computer Laboratory, 15 J.J. Thomson Avenue,
Cambridge, CB3 0FD, U.K.
Austin.Donnelly@cl.cam.ac.uk

**Abstract.** Active Network nodes are increasingly being used for non-trivial processing of data streams. These complex network applications typically benefit from protection between their components for fault-tolerance or security. However, fine-grained memory protection introduces bottlenecks in communication among components. This paper describes memory protection in Expert, an OS for programmable network elements which re-examines thread tunnelling as a way of allowing these complex applications to be split over multiple protection domains. We argue that previous problems with tunnelling are symptoms of overly general designs, and we demonstrate a minimal domain-crossing primitive which nevertheless achieves the majority of benefits possible from tunnelling.

## 1. Introduction

Modern network elements have many software components – for instance to support user-programmability. Software systems in such contexts must address a fundamental concern, that of multiplexing (sharing) the network element amongst many users. Generally, this involves some form of sandboxing to allow code to be executed on behalf of untrusted users [19, 16]. Consequentially, the multiplexing scheme must trade off between performance and security as well as other factors.

Sandboxing can be performed either at the language level (by using safe languages such at Java or ML), or at the machine level (by appropriate memory protection and CPU features). There has been much prior work on language-level sandboxing [1, 15, 5], and most current EEs (Execution Environments) rely on these techniques [21, 13]. However, language-level sandboxing lacks flexibility: invocations between components written in different languages must negotiate some common data marshalling format. Language-level sandboxing also reduces the utility of large bodies of pre-existing code by making it harder to re-use them.

Using hardware facilities to control access to memory and schedule the node's resources is desirable because it allows any language to be used, permitting legacy code re-use. Marshalling can be efficient since native machine formats for data can be used. The main drawback of using hardware to protect the EEs is that

J. Sterbenz et al. (Eds.): IWAN 2002, LNCS 2546, pp. 48–61, 2002.
© Springer-Verlag Berlin Heidelberg 2002

communication between protection domains is expensive due to context switches and cache invalidations. Furthermore, these penalties are increasing: as CPU speeds rise, more and more of their performance comes from effective caching of memory contents, branch predictions, and speculative execution. Frequent switching makes these caches ineffective.

Overall, these costs dissuade application designers from placing their modules in separate protection domains, especially if there is a continual stream of data passing through the application.

Thompson wrote: "The UNIX kernel is an I/O multiplexer more than a complete operating system. This is as it should be." [20]. This vision of a simple I/O multiplexer is one to which we find ourselves drawn once again, this time in the context of Active Network nodes. We introduce Expert, an OS designed specifically for network elements, filling this I/O multiplexer niche [4]. In this paper, we describe how Expert's memory protection architecture and its lightweight thread tunnelling directly support modular hardware-protected applications without suffering an undue performance penalty.

A good multiplexer will schedule the resource it manages. To this end, Expert uses the concept of a *path* (first introduced in the Scout OS [14]) to represent a flow of packets and their associated processing resources. The alternative of using multiple processes chained into a pipeline causes several problems: (1) extra context switching adds overhead; (2) as each process has its own scheduling parameters, it only takes one under-provisioned process to rate-limit the entire pipeline; (3) per-flow resource reclamation is complex, needing all processes to participate, and atomic revocation may be impossible; and (4) if multiple flows with different service characteristics are to be processed by the pipeline then each process must ensure it sub-schedules internally.

Unlike Scout, Expert paths can seamlessly cross protection domains. As an example, Fig. 1 shows the execution trace of a path which has tunnelled from module A to execute privileged code in module B, which in turned tunnelled into module C before returning back to module B and thence to A. Modules A, B, and C are all in separate protection domains, allowing B and C to implement trusted functionality securely.

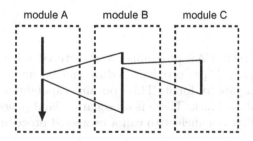

**Fig. 1.** Tunnelling between protection domains in Expert.

Section 2 discusses existing thread tunnelling schemes, and their shortcomings. Section 3 describes Expert's lightweight thread tunnelling primitive, and how it allows protected modules to be entered. A transcoder used as an example application is described in Sect. 4, and results quantifying the performance of the tunnelling system and the transcoder built over it are presented in Sect. 5. Section 6 concludes this paper and suggests areas for further work.

## 2. Background

Thread tunnelling was originally proposed as a solution to the performance problems observed in micro-kernel systems, where much inter-component communication takes place. In this guise, tunnelling is usually integrated into the IPC mechanism [2, 11], rather than using a message-passing approach.

Thread tunnelling designs all need to perform a number of core functions. A thread tunnelling primitive takes a thread and changes its runtime environment without passing through the scheduler: how much is changed depends on the individual design. At the very least, the thread's memory access rights are changed; multiple address-space operating systems may also need to switch address space. The vast majority of systems integrate a procedure call (i.e. a program counter change) with the tunnelling primitive. Most systems switch to an alternate stack, but there is some variety in how and when this stack is allocated. Asynchronous events may be masked while a tunnelled call is in progress. Table 1 compares five tunnelling systems against a selection of these features.

**Table 1.** Thread environment changes in different tunnelling systems.

| OS | rights change? | PC forced? | addr. space switch? | stack switch? |
|---|---|---|---|---|
| The CAP | ✔ | ✔ | ✘ | ✔ |
| Spring | ✔ | ✔ | ✔ | ✔ |
| Mach | ✔ | ✔ | ✔ | ✔ |
| Escort | ✔ | ✘ | ✘ | ✔ |
| Expert | ✔ | ✔ | ✘ | ✘ |

**The CAP.** Cambridge CAP programs are structured using a number of *protected procedures* [22]. A protected procedure can be invoked by any process having an *Enter* capability for it. This contains capabilities for the protected procedure's code, data, stack. There is no separate bind phase; mere possession of an Enter capability is sufficient to call a protected procedure.

**Spring Shuttles.** Spring introduces *doors* and *shuttles* as the building blocks of its IPC system [7]. Spring *doors* are capabilities, possession of which allows a call to be made to another protection domain. Each door is tailored to a specific client at bind time to allow servers to track their callers. When a call is made through a door, a free server thread is selected by the kernel, and resumed at the door's entry point; this creates a chain of threads, one per server called. Resource accounting and scheduling information is kept in a *shuttle* object shared by all threads in such a call chain.

**Mach Migrating Threads.** Ford and Lepreau [6] modified the Mach 3.0 IPC system to add tunnelling behaviour. They call their new scheme *migrating threads*. While they quote impressive speedups (a factor of 3.4 improvement when using migrating threads in place of static threads), they encounter practical problems: the need to support thread debugging/tracing, and propagating aborts due to server failures.

**Paths Crossing Protection Domains.** The designers of Scout [14] argued convincingly that paths are a good way of encapsulating the scheduling and resources used by flows of packets traversing a system. In Scout, data travels between modules along pathways determined at connection setup time. Escort [18] refines Scout by allowing protection domain boundaries to be specified, thus allowing modules with similar trust requirements to share memory protection rights while being protected from other modules in the system. However, protection crossings in Escort are expensive and Escort does not allow batching of packets before pushing them across a protection boundary, so it is impossible to amortise the cost of a protection switch over multiple work units.

## 3. Thread Tunnelling in Expert

Expert starts from the premise that the thread tunnelling primitive should be as simple as possible while still being flexible enough to support more complex schemes. The tunnelling primitive only changes memory access rights and forces the program counter to the module's entry point – state switching and other environmental modifications are delegated to the called code. We now describe Expert's thread tunnelling architecture in detail.

### 3.1. Pods

Expert binds protection domains (pdoms) to protected code modules (pods). Each pod's code is only executable by the pod's associated (or *boost*) pdom; paths executing within the pod run in this protection domain. Paths can trap into the pod by invoking a special kernel call which notes the boost pdom and forces the program counter to the pod's advertised entry point.

The kernel keeps a stack recording the boost pdom of each pod called, allowing nested invocation of one pod from another. The access rights in force at any time are the union of all pdoms on the boost stack plus the current base pdom.

This leads to a "concentric rings" model of protection. It makes calling into a pod a lightweight operation because the access rights are guaranteed to be a superset of those previously in force, so no caches need to be flushed on pod entry. Returning from a nested call is more expensive: there is by definition a reduction in privileges, so over-privileged data must be flushed from the TLB (Translation Lookaside Buffer) and other caches. However, this penalty is unavoidable; it is mitigated in systems which permit multiple nested calls to return directly to a prior caller (e.g. EROS [17]) but we believe that common programming idioms mean that such tail-calls are rare. For example, error recovery code needs to check the return status from invoking a nested pod; often a return code needs mapping.

All pods are passive: they have no thread of execution associated with them. This is in contrast with most other thread tunnelling schemes where the servers tunnelled into can also have private threads. Passive pods make recovery from memory faults easier since they are localised to the tunnelled thread.

## 3.2. Binding

Before allowing client threads to make pod calls, Expert requires them to bind to a pod offer. This enables bind-time access control checks and per-client initialisation. A pointer to this per-client state is stored with the binding record and supplied to the pod on each call. As usual, explicit binds allow a pod to be instantiated multiple times, allowing clients to select between instances by binding to a specific pod offer.

Expert uses a two-phase binding scheme illustrated in Fig. 2. First, clients contact a privileged Pod Binder process (stage 1.1) and ask it to setup an initial binding to a named pod offer (stage 1.2). The initial call on this temporary binding is special: it is interpreted by the pod as a bind request (stage 2.1), and may fail if the pod chooses not to accept this client. Otherwise, the temporary binding is upgraded to a full binding (stage 2.3). This two-phase scheme allows the pod to initialise any per-client state with full access to the calling client's particulars.

## 3.3. Invocation

Expert introduces a call_pod() system call, which takes as arguments a binding ID and an opaque pointer to the arguments.

**Pre-switch Checks.** call_pod() checks that the binding ID is within the client's binding table, and that there is room on the boost pdom stack for another pdom; these are the only tests needed, and both are simple comparisons of a value against a limit. The binding ID is used to index into the client's binding table to recover the pod to be called, and the state pointer to be passed in. Invalid (i.e. unallocated) binding IDs are dealt with by having the Pod Binder pre-allocate all possible binding IDs to point to a stub which returns an error code, thus creating a fast path by eliminating such error checking from the in-kernel code.

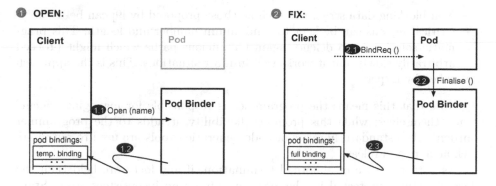

**Fig. 2.** Two-phase binding in Expert.

**Argument Marshalling.** There is no marshalling needed when a client tunnels into a pod – the native machine conventions are used. Large arguments are passed by reference in pre-negotiated external buffers. Of course, code in the pod must check that pointers received from the client identify state which is valid for that caller.

**Pdom Stack Manipulation.** The final step before calling the pod is to push the boost pdom onto the pdom stack, thus making available the rights granted by that pdom the next time a page fault occurs. When the pod call returns the boost pdom is popped off the pdom stack, page table entries are modified, and the TLB is flushed. This ensures that pages which were previously accessible while running in the pod are no longer available. To achieve this, the kernel arranges for control to return to itself after the pod call, rather than directly to the user application.

### 3.4. Concurrency Control

Taking out locks in pods can cause an effect similar to priority inversion: paths tunnelling into a pod bring their resource guarantees with them, so paths with a large CPU guarantee can be stalled by a path with a lower CPU guarantee. Several solutions exist:

- Paths running in critical regions can inherit the highest CPU guarantee of all blocked paths. This is analogous to priority inheritance in a priority-based system [10].
- Servers could provide "extra" cycles to ensure a minimum rate of progress through critical sections, or servers could simply reject calls from threads with insufficient CPU guarantees [12].
- Blocked paths could donate their cycles to the path currently running in the critical region, thus "pushing" it through. The pushed path later repays the loaned cycles once it has exited the critical section [8].

- Non-blocking data structures such as those proposed by [9] can be used.
- Critical regions can be kept to a minimum number and length. This pragmatic solution is no defence against malicious paths which might proceed arbitrarily slowly, but it works well in most situations. This is the approach taken by Expert.

Note that this means the programmer is responsible for managing concurrency themselves; while this promotes flexibility, it adds to the programmer burden unless standard libraries or code-generation tools are used to automate lock acquire and release.

As with any case of abnormal termination, if any locks are held then the data structures protected by the locks may be in an inconsistent state. Standard solutions to this include rolling back changes, working on shadow copies before committing, forced failure of the whole component, or using lock-free data structures.

### 3.5. Stack Switching

Unlike Expert, most thread tunnelling systems use a new stack for each protection domain. This prevents threads which have not tunnelled from manipulating the tunnelled thread's stack and/or snooping sensitive intermediate data.

However there are tantalising advantages to not switching stacks on protection switch. The performance is better, since arguments can be passed on the stack directly without copying. Stack pages are likely to be in the cache and to have a valid TLB entry. Also, memory usage is reduced by requiring only $T$ stacks rather than $T \times P$ for $T$ threads traversing $P$ pods.

Expert's `call_pod()` system call does not switch stacks, but instead uses the caller's stack while running inside a pod. The stack access rights are unmodified. This is safe because Expert does not allow other threads from the same base pdom to run while one is engaged in a tunnelled call. A pod may manually re-enable multi-threading if it deems it safe to do so (e.g. after having switched to another stack). The scheme has the twin merits of being simple and fast.

In any case, a paranoid pod can manually perform a stack switch as its first action, since the binding state passed in by the kernel can contain a pointer to a pre-prepared stack. If the kernel were to always switch stacks, a pod would no longer have the choice of whether to run on the same stack as its caller or not.

## 4. An Audio Transcoder

In this section we describe an example application which benefits from being decomposed into separate modules. We assume an Internet radio station which produces its output as a 44.1kHz stereo 192kbit/s stream of MPEG-1 Layer III audio (MP3). The illustrated application transcodes this source stream into three tiers: "gold" (the premium stream), "silver" (44.1kHz stereo, 128kbit/s, at a reduced price) and "bronze" (11kHz stereo, 32kbit/s, available for free). The transcoders are positioned on Active Network nodes close to the clients they serve, minimising the traffic crossing the core.

We describe the transcoder's design, and show how Expert allows precise control over the scheduling and protection of the various components. Control over resource scheduling allows the transcoder to degrade the level of service experienced by non-paying customers to ensure that paying customers are served promptly. The fine-grained protection offered by Expert should also increase the robustness of the system, although this effect is evidently hard to quantify.

### 4.1. Requirements

**Isolation.** Music fidelity should reflect payment. The listeners who have paid nothing must make do with whatever spare capacity is available in the system. Thus the gold, silver and bronze tiers are not only media quality metrics, but should also reflect the OS resources needed while processing streams of these tiers to ensure that streams from higher tiers are processed in a timely manner without loss.

**Per-client customisation.** Per-client customisation is needed, for example to target adverts, offer different disc-jockey "personæ", provide personalised news and weather, encryption or watermarking; we use AES (Advanced Encryption Standard) in our example.

**Protection.** Individual application components should be able to access only the areas of memory they need to perform their functions. For example, the encryption keys should only be readable by the encryption modules, so that key material cannot be leaked from the system.

### 4.2. Transcoder Architecture

Figure 3 shows how the transcoder application is segmented into components. Arrows depict packet movement and rate; thicker arrows correspond to higher data rates. The modules implementing the basic functionality are shown in rounded rectangles: DEC is an MP3 decoder instance, each ENC is an MP3 encoder instance, and each AES is an instance of an encryption module together with its key material. The rectangles represent the scheduled paths in this system: the common-rx path handles network receive, decoding, and encoding; the gold paths perform encryption and transmission, one per gold stream; the silver paths do the same for each silver stream; and the bronze path encodes and transmits one or more bronze streams. This diagram shows three streams at each tier. Note that unlike the others, there is a single bronze path to handle all free bronze clients; using a path per stream allows independent scheduler control over each of the paid-for streams to meet the isolation requirement, and provides memory protection.

The three caches shown are implemented as pods to allow sharing of their data, and the AES encryption modules are pods to minimise key material visibility within the application.

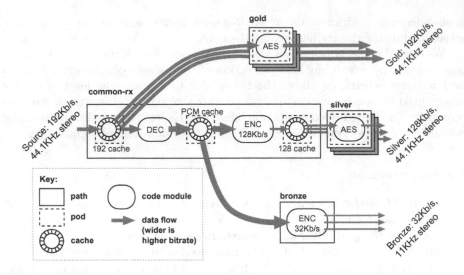

**Fig. 3.** Data flow through the transcoder. See text for description.

We implemented this architecture in two ways: a *path-based* variant, and an *all-in-one* variant using a single task containing multiple threads. The path-based version schedules the data flow through the system, and provides memory protection between components as described above. The all-in-one does neither: it is a baseline configuration to assess the overhead required to provide protection and scheduling.

## 5. Results

The test platform in all these experiments is an Intel Pentium Pro system running at 200MHz, with 32MB RAM, 256kB L2 and a split L1 cache: 8kB I / 8kB D. The test machine runs the Expert OS, and the transcoder application in either the path-based or all-in-one variant. It is easy to overload this modest machine: looking at systems when they are overloaded is instructive because this is where differences in architecture matter – if a system cannot shed load in a controlled fashion then it cannot offer different service levels, and is vulnerable to denial of service attacks.

### 5.1. Micro-benchmarks

Despite the obvious pitfalls of micro-benchmarks [3], they can be used to give a rough idea of the cost of various primitives.

Table 2 compares the cost of various protection switching schemes under Linux 2.2.16 and Expert. It shows how many cycles it takes to execute various types of call. Results for both hot and cold caches are given: "hot" is the average of 100,000 back-to-back calls; "cold" is the exponentially weighted moving average of 40 calls, with activity between each timed call to ensure the caches are filled with unrelated code and data. The cold cache number is more meaningful since (in an optimised system) calls which span protection domains are likely to be made infrequently and thus without already being in the cache.

**Table 2.** Cycles taken for different calls with hot and cold caches.

| OS | proc call | system call | pod call | pipe bounce |
|---|---|---|---|---|
| Linux (hot) | 7 | 340 | n/a | 3500 |
| Linux (cold) | 44 | 760 | n/a | 9100 |
| Expert (hot) | 7 | 280 | 2900 | n/a |
| Expert (cold) | 44 | 460 | 5000 | n/a |

The tests are as follows: "proc call" is a C-level procedure call to a function which takes no arguments and returns no value. The "system call" is `getpid()` on Linux, and a comparable minimal system call on Expert. The "pod call" is a switch from the untrusted client protection domain to a trusted pod environment and back again (the kernel implementation of `call_pod()` comes to just 44 instructions on Intel x86). The "pipe bounce" test sends 4 bytes to another process and waits for a response; this emulates the kind of lightweight IPC that pod calls can replace.

The table shows that an Expert pod call is between 17% and 45% faster than IPC on Linux. It also has better cache behaviour, as can be seen from the cold cache numbers. If the pod being entered is configured to switch to a private stack, the cost rises to 4100 / 6300 cycles for hot and cold caches respectively. Instrumenting the protection fault handler shows that this extra cost arises because twice as many faults are taken when the stack needs to be switched.

## 5.2. Transcoder Application

A Pentium II 300MHz running Linux is used as the "radio station" source, sending a stream of 192 kbit/s MP3 frames. Each frame (typically around 627 bytes) is encapsulated in a UDP packet and sent to the transcoder. The stream lasts around 146 seconds, and is paced to be delivered in real-time. Transcoded output is sent to a third machine which runs `tcpdump` to calculate the rates achieved by each stream.

**Cost of Protection.** In this experiment, the transcoder runs on an otherwise unloaded system. The amount of CPU time it consumes is recorded by the scheduler as a fraction of the total cycles available.

The transcoder application is initially configured for one gold, one silver and one bronze stream. We measure the CPU time required by both the path-based and the all-in-one variants as additional silver streams are added until the machine is saturated.

Table 3 shows the measured CPU time required for loss-free operation of the transcoder. The "path-based" row shows the cost for the path-based variant; the other row shows the cost for the all-in-one variant. The path-based version is, as expected, more expensive. Adding protection and proper scheduling costs between 2% and 6% only.

**Table 3.** Percentage CPU time required to service one gold, one bronze stream against a varying number of silver streams.

| | # silver streams | | | | | |
|---|---|---|---|---|---|---|
| Variant | 1 | 3 | 5 | 7 | 9 | 11 |
| path-based | 89 | 92 | 95 | 98 | – | – |
| all-in-one | 87 | 88 | 90 | 92 | 93 | 96 |

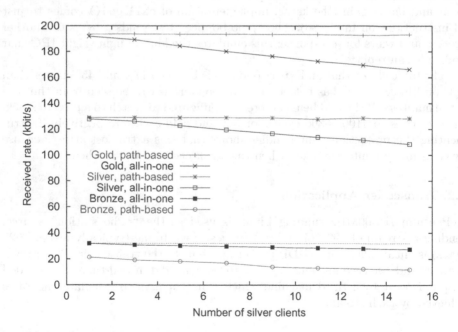

**Fig. 4.** Achieved rates for gold, silver, and bronze streams with and without isolation.

**Benefits of Isolation.** In this experiment, the transcoder services five gold streams, one bronze stream, and an increasing number of silver streams. The path-based version is configured to give the common-rx path a 45% share of CPU, the gold and silver paths each get 2%, and the bronze path 15% plus access to any "slack" time in the system. These guarantees are sufficient to meet the CPU needs for the common-rx, gold and silver paths, but the bronze path ideally needs approximately 46%. This means that the bronze path will mostly be running on slack time, i.e. as the number of silver paths increase, the CPU available to the bronze path will diminish. In this manner, the transcoder's administrator has expressed the policy that the bronze path's performance is unimportant compared to the common-rx, gold and silver paths.

For the all-in-one version of the transcoder, the task is allocated 85ms/100ms allowing it to monopolise almost all the machine's resources. The receiver records the average bandwidth achieved by an average gold (i.e. 192kbit/s) and silver (128kbit/s) stream, and the average bandwidth of the single bronze stream (ideally 32kbit/s).

Figure 4 shows these bandwidths both for the all-in-one and the path-based settings. Ideally, all the lines should be horizontal and co-incident, which would indicate that regardless of offered load, the streams continue uninterrupted. However, it is clear that the all-in-one design suffers large amounts of loss as the load increases. In comparison, the path-based gold and silver streams continue almost unhindered, all the losses being concentrated on the bronze stream.

# 6. Conclusion

We described how Expert binds protection domains to code modules and lets threads tunnel into these "pods", allowing fine-grained memory protection. Expert's efficient and flexible tunnelling support allows this protection to be used, even in the kinds of I/O-intensive applications typical of Active Network nodes. Expert's use of paths is well-suited to scheduling the resource consumption of such I/O-driven applications.

In an example application memory protection added a cost of between 2%-6%. The example also showed the benefits arising from correct scheduling of the CPU expended in processing media streams: under high load the CPU allocations of valuable streams were protected by sacrificing the performance of others. These features of Expert should make it an attractive base for running multiple Active Network Execution Environments.

## 6.1. Future Work

Nested pod invocations set up concentric rings of protection; relaxing this requirement would allow more flexible security policies, but makes each call slightly more expensive. Investigating the overhead involved might be worthwhile.

As yet, no Execution Environments have been ported to Expert. The RCANE system [12] showed how standard EEs perform when run over an OS with direct support for quality of service and tight memory protection; we expect similar results for Expert, with the additional benefit of being language-neutral.

Current work on Expert focuses on the uni-processor case; however work-station-based routers of the future are likely to have either multiple symmetric CPUs or a hierarchy of CPUs at a variety of distances from the data-path. Extending Expert to these multi-CPU machine architectures is an obvious step.

## 6.2. Acknowledgements

I would like to thank Jonathan Smith at the University of Pennsylvania, whose encouragement helped make this paper happen. I would also like to thank Tim Harris, Keir Fraser, and the anonymous reviewers for their helpful feedback.

# References

[1] Brian Bershad, Stefan Savage, Przemyslaw Pardyak, Emin Gun Sirer, David Becker, Marc Fiuczynski, Craig Chambers, and Susan Eggers. Extensibility, Safety and Performance in the SPIN Operating System. In *Proceedings of the 15th ACM Symposium on Operating System Principles (SOSP-15)*, pages 267–284, Colorado, December 1995.

[2] Brian N. Bershad, Thomas E. Anderson, Edward D. Lazowska, and Henry M. Levy. Lightweight remote procedure call. *ACM Transactions on Computer Systems*, 8(1):37–55, February 1990.

[3] Brian N. Bershad, Richard P. Draves, and Alessandro Forin. Using microbenchmarks to evaluate system performance. *Proceedings of the Third Workshop on Workstation Operating Systems*, pages 148–153, April 1992.

[4] Austin Donnelly. *Resource Control in Network Elements*. PhD thesis, Cambridge University Computer Laboratory, January 2002. Also available as CUCL Tech. Rep. 534.

[5] Marc E. Fiuczynski, Richard P. Martin, Tsutomu Owa, and Brian N. Bershad. Spine: A safe programmable and integrated network environment. In *Proceedings of Eighth ACM SIGOPS European Workshop*, September 1998. See also extended version published as University of Washington TR-98-08-01.

[6] Bryan Ford and Jay Lepreau. Evolving Mach 3.0 to a migrating thread model. In *Proceedings of the 1994 Winter USENIX Conference*, pages 97–114, January 1994.

[7] Graham Hamilton and Panos Kougiouris. The Spring nucleus: A microkernel for objects. In *Proceedings of the USENIX Summer Conference*, pages 147–159, Cincinnati, OH, June 1993.

[8] Timothy L. Harris. *Extensible virtual machines*. PhD thesis, Computer Science Department, University of Cambridge, April 2001. Also available as CUCL Tech. Rep. 525.

[9] Timothy L. Harris, Keir Fraser, and Ian A. Pratt. A practical multi-word compare-and-swap operation. *Proceedings of the 16th International Symposium on Distributed Computing (DISC 2002)*, 2002.

[10] B. W. Lampson and D. D. Redell. Experience with processes and monitors in Mesa. *Communications of the ACM*, 23(2):105–117, February 1980.

[11] Jochen Liedtke. Improving IPC by kernel design. In *Proceedings of the 14th ACM Symposium on Operating System Principles (SOSP-14)*, pages 175–188, Asheville, NC, December 1993.

[12] Paul Menage. RCANE: A Resource Controlled Framework for Active Network Services. In *Proceedings of the First International Working Conference on Active Networks (IWAN '99)*, volume 1653, pages 25–36. Springer-Verlag, 1999.

[13] Jonathan T. Moore, Michael Hicks, and Scott Nettles. Practical programmable packets. In *Proceedings of the 20th Annual Joint Conference of the IEEE Computer and Communications Societies (INFOCOM'01)*, April 2001.

[14] David Mosberger and Larry L. Peterson. Making paths explicit in the Scout operating system. In *Proceedings of the 2nd Symposium on Operating Systems Design and Implementation (OSDI'96)*, pages 153–167, Seattle, Washington, October 1996.

[15] George C. Necula. Proof-carrying code. In *Proceedings of the 24th ACM SIGPLAN-SIGACT Symposium on Principles of Programming Languages (POPL '97)*, pages 106–119, Paris, January 1997.

[16] Larry L. Peterson, Scott C. Karlin, and Kai Li. OS support for general-purpose routers. In *Proceedings of the 7th Workshop on Hot Topics in Operating Systems (HotOS-VII)*, pages 38–43, March 1999.

[17] Jonathan S. Shapiro, David J. Farber, and Jonathan M. Smith. The measured performance of a fast local IPC. In *Proceedings of the 5th International Workshop on Object Orientation in Operating Systems*, pages 89–94, Seattle, WA, November 1996.

[18] Oliver Spatscheck and Larry L. Petersen. Defending against denial of service attacks in Scout. In *Proceedings of the 3rd Symposium on Operating Systems Design and Implementation (OSDI'99)*, New Orleans, Louisiana, February 1999.

[19] David L. Tennenhouse and David J. Wetherall. Towards an active network architecture. *ACM Computer Communications Review (CCR)*, 26(2):5–18, April 1996.

[20] Ken Thompson. Unix implementation. *Bell System Technical Journal*, 57(6, Part 2):1931–1946, July/August 1978.

[21] David J. Wetherall, John V. Guttag, and David L. Tennenhouse. ANTS: A toolkit for building and dynamically deploying network protocols. In *Proceedings of the 1st IEEE Conference on Open Architectures and Network Programming (OPE-NARCH '98)*, April 1998.

[22] Maurice V. Wilkes and Roger M. Needham. *The Cambridge CAP computer and its operating system*. Elsevier North Holland, 52 Vanderbilt Avenue, New York, 1979.

# RADAR: Ring-Based Adaptive Discovery of Active Neighbour Routers

Sylvain Martin and Guy Leduc

Research Unit in Networking, Université de Liège, Institut Montefiore B28, 4000
Liège 1, Belgium
{martin, leduc}@run.montefiore.ulg.ac.be
http://www.run.montefiore.ulg.ac.be/

**Abstract** The RADAR protocol and its underlying neighbourhood discovery framework extend the ANTS toolkit by giving active nodes the
ability to discover dynamically other active nodes close to them without
relying on any configuration file. Such an automatic discovery is the key
to administration of large or sparse active networks and the first step
towards an efficient active routing.

Active nodes will use their local IP routing table to run an extended ring
search in their domain. An Additive Increase Multiplicative Decrease
control allows RADAR to discover several neighbours per physical interface without searching too far away or fixing a maximum distance a
priori. The protocol is complemented by a traffic-driven discovery that
can grab capsules coming from unknown nodes (mainly outside the local
domain) and trigger targetted probing of those addresses.

## 1 Introduction

### 1.1 Purpose of This Work

Active networks often require the network engineer to develop his own active
routing that would better fit the application needs than what a default routing
protocol would have done. One can, for instance, use active nodes *soft state* so
that a multicast flow of capsules reaches all of its subscribers by simply following
indications that have been stored by previous *subscribe capsules* (that follow the
'upstream').

However, to have such schemes work, we need an initial way to find routes
to nodes. Clearly, in the previous example, the *subscribe capsules* cannot rely on
some soft state to reach the stream source. We have different ways to deal with
that kind of capsules :

1. if we know the stream source's IP address, we can rely on IP routing to reach
   the source.
2. if we know the stream source's active address, we could have an *active routing
   protocol* that would relay the capsules to the *next active hop* towards the
   stream source.

J. Sterbenz et al. (Eds.): IWAN 2002, LNCS 2546, pp. 62–73, 2002.

3. otherwise, we could forward the subscribe capsule to all the possible active next hop repeatedly until we find the source (that's awfully unscalable, though).

The first option is by far the simplest if its condition is met. However, it would lead to create a direct tunnel between the source and the subscriber, which prevents intermediate nodes from doing any processing on the capsules.

In both other options, we need to know a list of *next active hops* of the active node, either to exchange active-routing information, or to select one and forward it directly the data. Those active nodes will be referred to as *active neighbours* or simply *neighbours* in this paper. Most execution environments assume either that the whole network is made of active routers or that the list of active neighbours is provided through some configuration file. This limits the use of these environments to relatively small or very static testbed networks.

So the purpose here is to design a *neighbours discovery protocol* for the ANTS[1] framework that will let every active node *dynamically* discover its active neighbours to form an *overlay network* on top of the IP topology and possibly inform an active routing protocol of neighbour arrival or departure. Moreover, we want it to be fully plug-and-play and require no centralized component.

## 1.2 Defining Neighbourhood

Because active nodes run on top of IP, every active node can virtually reach any other active node. However, we want to reduce this "full mesh" to an overlay topology that better matches the real (physical) topology. Our protocol will perform that transformation by selecting a list of *neighbours* among all the reachable active nodes.

Active Neighbours are the equivalent of the next hops in IP routing and have to meet the following conditions:

- They must be active nodes,
- They must be directly reachable by the current node, which means if the current active node sends an IP packet to a neighbour, no other active router will receive the packet before it has reached the neighbour,
- They must be 'close enough' to the current node.

The latter condition enforces that a restricted amount of resource will be consumed to reach a neighbour node. For instance, we could fix a maximum TTL needed to reach a neighbour or define a time interval for the neighbour to respond. Note that the neighbourhood relationship is not necessarily symmetric, especially when routes aren't (i.e. if the route leading from $A$ to $B$ is not the route leading from $B$ to $A$).

## 1.3 Structure of This Paper

After the state of the art summary in section 2, section 3 introduces the reader to the basic mechanisms that are used in our protocol.

The extended-ring-based discovery protocol - *RADAR* - is introduced and described in section 4. It is built on top of the framework of section 3, and adds an efficient capsules generation policy.

The architectural changes that have been made to the ANTS toolkit to make our protocol run on it are described in section 5.

## 2   State of the Art

Neighbourhood discovery techniques in overlay networks significantly differ from those in classical or ad-hoc networks by the fact that we cannot reach all the potential neighbours by just broadcasting a message through an interface. Although it re-uses the principle of the *extending ring search* [14], *RADAR* doesn't require any node to support multicast and works even without broadcasting facility.

Existing active network Execution Environments rarely address the problem of active neighbours discovery. In ANTS [1] and PLAN [4], routing tables and neighbourhood tables are read out of static configuration files. Using the DANTE [5] protocol from ABone [6] specifications, a single domain can join a statically configured backbone router and start exchanging routing updates, but these features don't help for auto-configuration of a dynamic overlay topology.

Other works such as the PROTEAN [7,8] project and its SPINE network infrastructure or the ALAN [9] project are based on a hierarchical database where all service providers will register. The hierarchy is usually based on domain names hierarchy and allows identification of which service is provided by which node. However, such databases don't provide a *neighbourhood* relationship and have virtually no information about components physical proximity [1]. So we can use them when a client has to find out which server it can connect to, but they aren't of much help when building dynamic active routing tables.

In the TAO dynamic overlay management algorithm [12], the neighbourhood discovery is based on DNS queries for a well-known name that returns a list of *cluster heads*, and each node uses ping statistics to join one of these clusters. Every node of such a cluster then becomes the neighbour of every other node in its cluster and an elected "cluster head" runs a routing protocol with peer clusters to build neighbourhood between clusters. Unlike what happens in our protocol, the TAO algorithm works at a coarse level: two nodes of a cluster can be neighbours even if they can't directly reach each other. Moreover, two nodes that are physically close to each other could use an excessively long path to communicate if they join distinct clusters.

## 3   Neighbourhood Discovery Framework

### 3.1   The AYA Capsules

The whole discovery process is based on very simple AYA (stands for Are You Active?) capsules - a sort of active *ICMP Echo* packet. Those capsules are sent

---

[1] unless assuming that a domain name is made of close devices, which is certainly false for domains as *.com*

to the node or network we want to probe and will be intercepted by the first active node they cross (see listing 1.1 for a pseudocode for AYA.evaluate()).

Only active nodes will recognize AYA capsules and reply to them. If such a capsule is received by a non-active IP node, this one will have no upper protocol that matches the ANEP protocol type written in the IP header and will just drop it (or possibly reply with an ICMP error message). Non-active intermediate nodes will just forward the capsule as they would forward any other IP packet.

Once the AYA capsule is received by an active node, this node will fill in the "neighbour" field with its own active address. Then the capsule stops searching its target and goes back home. The neighbour discovery application running on that home node will then learn that it has a (new) neighbour, but also that this neighbour is on the road to reach the target address.

**Listing 1.1.** evaluate() method for AyaCapsule

```
vars :
neighbour = NOT_FOUND;
target; /* probed address */
evaluate on node N:
  if node address = capsule source {
    if neighbour != NOT_FOUND
      "deliver to RADAR instance";
    else
      "route to target address";
  } else {
    if neighbour = NOT_FOUND
      neighbour= node address;
    "route back to capsule source";
  }
```

**Listing 1.2.** Pseudocode for RADAR probing policy

```
on every interface:
  repeat {
    testing = targets.nextRing();
    foreach (1..3) {
      testing.sendCapsules();
      wait(DELAY);
    }
    threshold+=testing.size;
  } until threshold.reached
    || testing.isEmpty;
  on "capsule c delivered" {
    "remove c.target from testing";
    threshold*=ALPHA;
  }
```

## 3.2 Neighbour Discovery as an Active Application

Rather than implementing the neighbour discovery protocol directly in the core of the ANTS execution environment, we decided to develop it as an *active application* that runs on top of that environment and uses it to send its capsules.

However, the neighbour-discovery application must have a privileged status in the ANTS execution environment to be able to do its job properly. For instance, it must be able to provide the enumeration of Neighbour nodes instead of the primordial node[2], and have the opportunity to inform the ANTS's routing table manager of neighbours arrivals and departures.

Special care should be taken while updating the ANTS environment to support extensions required by the neighbourhood discovery, so that appropriate security checks are performed and only "trusted" applications can modify the neighbourhood table.

---

[2] In ANTS, the PrimordialNode is the main component of the Execution Environment that deals with most communication with the NodeOS, including retrieving routing and neighbourhood tables

## 3.3   Capsules Grabbing

A protocol like the neighbour discovery protocol[3] can't work using only the regular capsule-over-udp scheme. In that scheme, a capsule is routed along a list of *udp tunnels* linking active nodes, and only the tunnel endpoints can do some processing on the capsule. This can be useful when trying to establish a service that only requires some of the network nodes (even if the whole network is made of active nodes) like transcoders or repeaters, but it completely ruins our plans when we're dealing with neighbour discovery.

Keeping this UDP service would mean that we'd have to send AYA capsules to every router on the path to a destination $D$ for discovering the neighbour which leads to $D$. Indeed, sending an AYA capsule with $D$ as IP destination would skip all intermediate active nodes and make $D$ appear as a neighbour if it is active, which is far from what we expected.

To have things working, we need active nodes to catch the AYA capsule and evaluate them *even if they're not the IP destination of this capsule*. This is what we call *capsules grabbing*. With a NodeOS-based system [11], this means that some ANTS entity should create a new InputChannel that will inspect the IP packet header and accept them if they have the right *protocol type*, regardless of the IP destination they hold. Once such packets are received on that channel, they will be processed by ANTS as if they were coming from another channel: retrieval of the capsule class based on its Method IDentifier and then capsule evaluation. Moving to this scheme might require an update of the active environment, but it doesn't require a modification of the NodeOS itself.

As we want to keep backward compatibility with the UDP tunnel model, we don't want to grab any kind of capsule, but only those that 'require' or 'request' it. In this work, we decided to use *grabbable* capsules every time a node is requested to send a capsule to a destination for which it has no neighbour. This gives us a reliable "escape route" (falling back to IP route table) and prevents the nodes from dropping capsules because of incomplete routing or neighbourhood tables.

The capsules grabbing is mandatory for our solution, but it isn't sufficient to solve the neighbourhood problem:

1. The active nodes that will perform operations along the route might not be right *on* the default IP route and therefore would not grab the capsules they should process.
2. To be grabbable, the capsule's address must contain (or map to) a unique IP address. For capsules that don't know their destination a priori this will not be possible.
3. Only capsules whose target address can be mapped to a unique IP address can be made grabbable. Having all capsules grabbable would require that active addresses (used for ANTS active routing) *are* IP addresses.

---

[3] In the rest of this paper, we will give the generic name of *Neighbour Discovery Protocol*(NDP) to any software component that extends ANTS to support a dynamically-built neighbourhood table. Our RADAR protocol is one possible instance of the abstract NDP protocol.

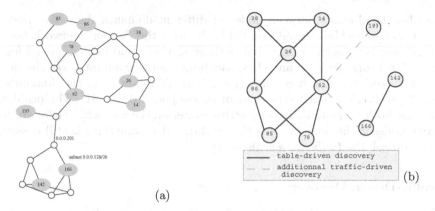

**Fig. 1.** *(a) topology used to test discovery protocol and (b) resulting neighbourhood mesh when table-driven discovery is used alone*

### 3.4  When to Send AYA Capsules ?

There are two possible approaches for the Neighbours Discovery Protocol: the *table–driven discovery* and the *traffic-driven discovery*. Better results are obtained when both discovery techniques are used together.

The table-driven approach is based on inspection of the underlying network-layer (IP) routing table, while traffic-driven discovery will try and learn from any protocol's capsules that the node receives or grabs.

**Table-Driven Discovery**

Every time a new entry is added in the routing table, the neighbourhood daemon is notified and will try to determine whether the *next hop* and the *destination* of this entry are active or not. When the Neighbour Application initializes itself, it will ask the *NodeOS* for a copy of the IP routing table and simply consider all the route table entries as 'new' and will process them for neighbour discovery.

In order to avoid the risk of high AYA traffic when a node boots up, route table entries are grouped by output interface and only one AYA capsule can be sent through an interface during a probing interval.

Figure 1a presents one of the topologies used for discovery testing, with active nodes highlighted. This network was in fact split into two independent domains (running shortest path algorithms) statically linked by a "backbone" node. No routing information about the domains internal routes crossed the backbone, as usual in hierarchical routing. Host 193 does not participate in any IP routing protocol and simply sends all its datagrams to its default router (the 'backbone').

As we can see on figure 1b, if the time to live (TTL) for the AYA capsules is correctly set, the discovery works quite fine: nodes within an area have found each other and the topology did not degenerate into a full-mesh. However, no neighbourhood exists between nodes of different domains. An active node located on the backbone would probably discover the "upper" network because its gateway is an active router, but nothing similar could be guaranteed for the "lower" network. As such an active backbone router wouldn't know the details of the "lower" domain, it could only send an AYA capsule to the domain's address[4] (0.0.0.128/26) or to the domain access point (0.0.0.201). The problem is that, for both of them, there is no active router on the way while there are some active nodes in the domain. This clearly shows that something is still missing in our protocol: the *traffic-driven* discovery.

**Traffic-Driven Discovery**

First, we can observe that this lack of active neighbourhood doesn't completely break the reachability of the whole network. Indeed, for each destination that has no active neighbour discovered, the active router will simply forward the capsule to the final destination using a *grabbable* IP packet. Moreover, the default route is forced to have no neighbour. Thus, for capsules that rely on the active routing table to reach their destination, it's still possible to reach all active nodes, even if some of them appear to be unreachable in the active topology.

The extension of our protocol will consist of using such capsules to dynamically *learn* new neighbours. Every time a capsule is received by an active node, the node's neighbour discoverer will lookup the *previous node* address stored in the capsule in its internal tables and will try to determine to which of the following categories it belongs:

- The previous node is an already known neighbour,
- The previous node has already been tested and is not a neighbour (e.g. too far or routes from and to that node are distinct),
- The previous node hasn't been tested yet.

In the latter case, we will enqueue the previous node's address in the incoming interface's list of potential neighbours and send it AYA capsules when its turn has come.

### 3.5   Protocol Behaviour Summary

The Neighbour Discovery Protocol is designed for quick configuration of ANTS nodes. It probes the network with AYA capsules to discover which nodes are active using its IP routing table or other capsules to guess potential neighbours.

- Unlike most existing techniques, it doesn't require configuration files on each node and is completely "plug-and-play", but it needs a running "daemon" on every active node to send AYA capsules periodically.

---

[4]  One of the characteristics of our discovery protocol is that it may send packets towards *network* addresses (found in the routing table) in order to reach the first active neighbour towards a given destination.

- Active nodes that don't have that daemon can still be discovered as neighbours, but can't discover other nodes.
- It can deal with neighbours crossing domain bounds if some data traffic goes from one domain to another
- Capsules with an IP destination can be routed even before discovery is completed.

In order to implement our neighbourhood discovery framework, the NodeOS must be able to provide a copy of the IP table[5]. This table will have to include at least *destination*, *next hop* and *cost* fields. Moreover, we will need a way for the NodeOS to notify changes in the routing table to the NDP framework. Finally, the addressing scheme for active nodes must include IP addresses, but is not limited to IP addressing.

# 4   R.A.D.A.R.: A Ring Search Using AYA Capsules

The generic neighbourhood discovery framework presented in section 3 must be completed with a *probing policy* that will choose in which order AYA capsules for table-driven discovery will be sent and which entries of the initial routing table will/won't be used for discovery.

## 4.1   Using Routing Table Information

In each active node, an independent instance of the probing policy is running for each network interface. At node initialization, the addresses of the routing table are distributed to each policy instance based on the interface used to reach each address.

The amount of capsules needed to discover a given topology is a crucial point of the protocol: sending useless[6] AYA capsules will lead to bandwidth and CPU waste when the network becomes larger (and thus when there are more potential targets). By ordering targets in an extending ring search rather than keeping the routing table order, *RADAR* can save some unnecessary probes.

In order to achieve a good scalability, *RADAR* is restricted to a $n$-hops neighbourhood for its table-driven discovery. Any entry which has a *cost* beyond the MAX_HOPS parameter will be automatically discarded. This prevents the protocol from using excessive resources on large networks with few active nodes. However, in sparse overlays, scanning the whole $n$-hops neighbourhood is already too costly, so we'll try to keep the MAX_HOPS only as a last resort limit.

## 4.2   Semi-persistent Searching

One simple way to reduce the discovery overhead is to stop searching if the physical neighbour is active. *RADAR* tries to extend this rule to the whole

---

[5] regardless of the protocol that built that table

[6] either AYA request duplicates asking for information we already have, or capsules trying to reach a target that is obviously too far for their limited resources.

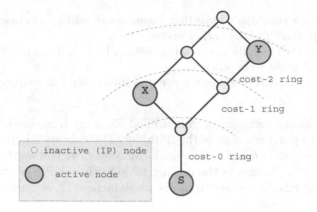

**Fig. 2.** *Example of impact of neighbourhood on routing performance*

discovery. Rather than stopping at the first neighbour it encounters, it has a half-persistent behaviour that makes it look a bit further and try to find more neighbours than just one per interface.

Figure 2 shows the point of doing so: if we stop our neighbourhood search when $X$ is discovered, then routing capsules from $S$ to $Y$ will require 5 hops: 2 to reach $X$ (nearest neighbour) and then 3 other hops to go from $X$ to $Y$. If *RADAR*'s threshold is properly set up, we can make the discovery daemon continue and find both $X$ and $Y$. Capsules will now be routable directly from $S$ to $Y$ without extra cost.

Restricting the search to the *cost ring* of the first discovered neighbour will often lead to sub-optimal routes, whereas a bit more probing overhead can easily discover better ones.

For each interface, a *threshold* value is maintained and defines how far the search will go on that interface. Technique used for correct threshold setup is presented at section 4.4

## 4.3   Tests Results

In order to validate our protocol, some tests have been issued on the brand-new *RUN Active Network Simulator* developed at Université of Liège - a cross between the legacy ANTS [1,2,3] toolkit and the generic SSFNET network simulator [10].

Compared to an earlier "naive" policy (testing every target in table order), *RADAR* is 6 times faster for a complete discovery on the topology shown in figure 1 and saves about 50% of the search overhead (only 234 Kbits of AYA capsules generated and forwarded against 439 in the naive approach), which lowers its cost roughly to the same level as capsule code download.

## 4.4   Setting the R.A.D.A.R. Discovery Threshold

The *threshold* value used to define how far[7] an interface must search is an important parameter for the whole protocol. It will define the trade off between search cost and routes efficiency, but also the 'complexity' of the discovered active topology. So a threshold-based limit allows us to set a less constraining absolute MAX_HOPS parameter, and lets the protocol search 'as far as it has to'.

Simulations have been carried out to find out a threshold definition algorithm that would be able to *adapt* to the underlying topology and that would achieve good resource use. Using only the cost of the first found neighbour to compute the threshold leads to a high dependence on parameters like active nodes density or spanning tree growth at each node.

## 4.5   Adaptive Threshold Setting Algorithm

The solution we adopted is based on the well known *additive-increase-multiplicative-decrease* generic algorithm widely used in network engineering. It tries to find a balance between discovered neighbours and potentially remaining neighbours.

Every time an AYA capsule successfully returns to an active node, the *threshold* of the sending interface is reduced multiplicatively by $\alpha$. Keeping from $1/2$ to $2/3$ of the previous value has given good results on every tested configuration (see figure 3). When all targets of the current *cost ring* have been probed, *threshold* is incremented by the amount of non-responding[8] targets. A pseudocode for *RADAR* behaviour is given in listing 1.2.

Note that after the first neighbour has been found for an interface (and has reduced *threshold*), the threshold reduction depends on how many targets that neighbour covers (i.e. every time such a covered target is probed, the neighbour replies and the threshold is further reduced). So, finding a neighbour located at the 'edge' of the network will not have much impact on neighbours search while finding a neighbour in the 'core' of the network quickly stops the search.

Note also that the threshold is only used when a ring search wasn't completely successful. If every target of a ring has replied, then the search won't go any further, regardless of the current threshold.

## 4.6   Using R.A.D.A.R. on Dynamic Topologies

In order to support active nodes arrival or departure at any time *RADAR* needs a timeout mechanism that will force it to re-check some targets (for nodes that are still active neighbours or that have been active in the past). We still use AYA capsule to refresh the information, but the *threshold* parameter will only be modified if the situation has changed since the last known state.

*RADAR* will send refresh capsules only if it has received no traffic from the target address, thus avoiding unnecessary overhead on heavily loaded links. In

---

[7] in hops count from the searching node
[8] after 3 re-emission of the AYA capsules with a period of DELAY

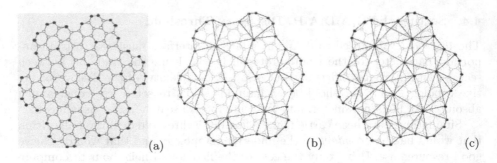

**Fig. 3.** *Discovery using adaptive (AIMD) threshold: (a) physical topology, (b)* $\alpha = 0, 5$, *(c)* $\alpha = 2/3$

addition, a *binary exponential back-off* mechanism will make the refresh attempts more and more spaced if the target does not send data capsules between AYA probes. By doing so, a node will usually wait a long time before discovering that a silent node is down while the crash of a router that exchanges a lot of data will be discovered almost immediately.

## 5  ANTS Modifications

### 5.1  Output Channels

The *RADAR* protocol requires some extensions to the legacy OutChannel class from ANTS, mainly to support *grabbable capsules* facility. A derivate output channel using ANEP/IP rather than ANEP/udp will be automatically generated when the next hop address can be recognized as an "unknown" address (i.e. when the requested route hasn't discovered any neighbour).

Moreover, it is now possible for OutChannels to become invalid due to a change in the ANTS routing table (done by a routing manager in response of some active neighbourhood notification)

### 5.2  Route Table Copy

*RADAR* requires a copy of the IP routing table for its discovery process. The *next hop* addresses of the table entries will be modified dynamically to reflect the topology knowledge of *RADAR*. The entries in this table will have to carry a *route cost* (preferably a hop count) as well as an identifier of the network interface card (using the next hop address is no more possible as it is now prone to dynamic updates).

## Conclusions and Future Work

We have extended the ANTS platform to support discovery protocols of active neighbours. The presented protocol - *RADAR* - performs that discovery with a cost comparable to code download when active nodes are dense or slightly sparse, but it could still be improved on very sparse topologies through a local *host cache service* like what is done in decentralized peer-2-peer networks [13].

On a single domain, *RADAR* almost always results in connected graphs[9]. When considering a hierarchical network, traffic-driven discovery is mandatory to interconnect subnets, and it will work better if border routers are active.

A feedback loop could be investigated to modify the heuristic $\alpha$ constant when the amount of discovered neighbours isn't satisfactory.

Thanks to the *traffic-based* discovery, any active application can remotely "guide" *RADAR* by contacting some specific targets (querying a directory of the domain's active node, using a multicast group, etc.). *RADAR* will then detect the new potential neighbours from reply-traffic analysis.

# References

1. D. Wetherall, A. Whitaker : ANTS - an Active Node Transfer System. version 2.0. *http://www.cs.washington.edu/research/networking/ants/*
2. D. Wetherall : Service Introduction in an Active Network. *http://www.cs.washington.edu/research/networking/ants/ants-thesis.ps.gz*
3. D. Wetherall, J. Guttag, D. Tennenhouse : ANTS - A Toolkit for Building and Dynamically Deploying Network Protocols. *IEEE OPENARCH'98*, April 1998
4. P. Kakkar : The Specification of PLAN (Packet Language for Active Networks) Draft 1, University of Pennsylvania (July 12, 1999)
5. S. Berson, B. Braden : DANTE : Dynamic Topology Extension for the ABone. *ABone: Technical Specs - http://www.isi.edu/abone/DOCUMENTS/dante2.ps*
6. S. Berson, B. Braden, L. Ricciulli : Introduction to the ABone. *http://www.isi.edu/abone/DOCUMENTS/ABoneIntro.pdf*
7. R. Sivakunnar, S.W. Han, Vaduvur Bharghavan : PROTEAN : A scalable Architecture for Active Networks *Proceedings of OPENARCH'2000*, March 2000
8. Raghupathy Sivakumar, Sungwon Ha, Sungwook Han, Vaduvur Bharghavan: The Protean Active Router: Design and Implementation. *The 14th IEEE Computer Communications Workshop (IEEE CCW'99), Invited Presentation*, October 1999.
9. A. Ghosh, M. Fry, J. Crowcroft : An Architecture for Application Layer Routing. *Lecture Notes in Computer Science 1942, "Active Networks", Springer, 2000 (IWAN 2000)*
10. Scalable Simulation Framework for modeling the Internet. *http://www.ssfnet.org*
11. Larry Peterson (Editor). NodeOS Interface Specification. *DARPA AN NodeOS Working Group Draft*, 1999.
12. Andy Collins, Ratul Mahajan, and Andrew Whitaker. The TAO Algorithm for Virtual Network Management. *Unpublished work*. December 1999. *http://citeseer.nj.nec.com/collins99tao.html*
13. Clip2. The Gnutella Protocol Specification v0.4 *http://www9.linewire.com-/developer/gnutella*
14. D.R. Boggs : Internet Broadcasting, *Ph. D. thesis, Electircal Engineering Dept., Stanford*, 1982 and *Technical Report CSL-83-3, Xerox PARC Palo Alto, California*

---

[9] The graph might be partitioned in some rare case, but traffic-driven discovery will usually reconnect them if some capsules cross the partition line

# Integrated Service Deployment for Active Networks

Matthias Bossardt[1], Takashi Egawa[2], Hideki Otsuki[3], and Bernhard Plattner[1]*

[1] Computer Engineering and Networks Laboratory, ETH Zürich, Switzerland
{bossardt|plattner}@tik.ee.ethz.ch
[2] NEC Networking Laboratories, 4-1-1 Miyazaki, Miyamae-ku, Kanagawa, 216-8555 Japan
t-egawa@ct.jp.nec.com
[3] Communications Research Laboratory, 4-2-1 Nukui-Kitamachi, Koganei-shi, Tokyo,
184-8795 Japan
otsuki@crl.go.jp

**Abstract** A key feature of active networks is the capability to dynamically deploy services. In this paper, we present a scheme to classify service deployment mechanisms of existing or future active network architectures. Distributed algorithms (services), as being implemented in active networks, can be described based on active packets or as distributed programs running on active nodes. Although both programming models are basically equivalent, some services are more naturally implemented in either way. This paper proposes an active node architecture that supports the implementation and deployment of services according to both programming models. We point out that a combination of in-band and out-of-band service deployment is needed to dynamically deploy services implemented in either model. Furthermore, we argue that composing services from service logic implemented in either programming model is beneficial for the design of efficient and flexible services. We reason that a service abstraction in the form of a service description language is necessary to cope with real world scenarios.

## 1 Introduction

The difficulties to deploy new services in IP networks led to the development of active networks. Historically, active networking technology was based on an active packet (or "capsule") model, where packets — in addition to the ordinary payload — carry program code, which is executed in appropriate execution environments (EEs) of active nodes [15].

Active packets have some advantages that are hardly achieved by other technologies. One merit is that the processing of each packet can be programmed separately. Another merit is that a certain class of services is easily implemented and *deployed*. Examples of such services include different flavors of multicast, congestion avoidance algorithms and others. The main advantage of active packets is that they do not require any service deployment infrastructure.

Research, however, showed also that active packets are not practical for some type of services [1]. Examples include services that interact with many packet flows, as is the

---

* This work is partly funded by ETH Zürich, and Swiss BBW under grant number 99.0533.

J. Sterbenz et al. (Eds.): IWAN 2002, LNCS 2546, pp. 74–86, 2002.

case for firewalls filters, or long lived, usually rather complex, services that provide control functionality (e.g. routing daemons). Moreover, the expressiveness of active packet programming languages is often restricted due to security or performance concerns. Thus, there is a clear need to extend an active packet based network with dynamically deployable service components.

In this paper, we analyze service deployment mechanisms for active networks according to a novel classification scheme. Based on this analysis, two complementary service deployment schemes are selected, which we propose to integrate in order to overcome the limited expressiveness of active packets and to support more generic services. Furthermore, we motivate a service abstraction in the form of a service description language. As a result, a flexible active network architecture is obtained.

The following sections are organized as follows. In section 2, the design space of service deployment in active networks is analyzed and existing active network systems are classified. Section 3 discusses an integrated approach to service deployment and motivates a service abstraction in the form of a service description language. Finally, section 4 concludes the paper.

## 2   Exploring the Service Deployment Design Space

The potential of active networks to provide a wide variety of services must be supported by an appropriate service deployment architecture. In fact, active networks are unmatched in their flexibility to accommodate new services, because they allow programmability not only in the management and control planes, as is the case for programmable networks, but also in the data plane. In this section, we explore the design space of service deployment and discuss different approaches found in the literature.

### 2.1   Two Levels of Service Deployment

Service deployment in a network can be subdivided into two levels:

1. The network level where nodes that run service components are identified.
2. The node level where software must be deployed within the node environment.

A comprehensive framework must support service deployment at both levels. Due to the loose coupling between the two levels, the design of service deployment mechanisms at each level can be tackled to a great extent independently. The design space is similar for both levels (cf. Figure 1). It is essentially defined by two logically orthogonal axes being described in the following sections.

### 2.2   In-Band versus Out-of-Band Service Deployment

The x-axis of the service deployment design space defines the way service logic is associated with certain nodes and distributed in the network. The designer has the choice between in-band and out-of-band service deployment. In-band deployment refers to a system, where the service logic is distributed in the same way as payload data. Out-of-band deployment, on the other hand, refers to an architecture where service deployment and payload data use logically and/or physically distinct communication channels. That is, service deployment information is exchanged via the control or management plane.

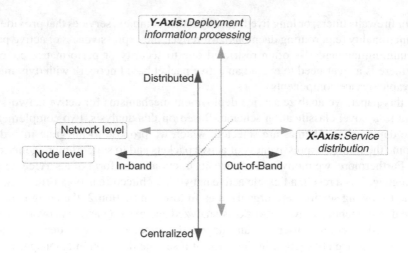

**Fig. 1.** Design space of service deployment frameworks

**2.2.1 Network Level.** Service deployment at the network level consists of identifying appropriate nodes that match the service requirements and are eligible to execute service logic.

In an in-band approach, service deployment is inextricably linked to the way packets are routed through the network. Only nodes a packet is routed to are eligible to execute service logic. In such a scheme, however, it may be possible to influence the location of service logic by executing/selecting different (e.g. service-specific) forwarding rules on the nodes. The main motivation for an in-band approach is that no specific service deployment infrastructure is needed. Active packets are a typical representative of network-level in-band service deployment.

In an out-of-band approach, on the other hand, node capabilities and topology information is gathered via management or control communication channels. Once the suitable nodes are identified, the same channels are used to allocate network level resources and to trigger the node level service deployment. Existing approaches to network-level out-of-band deployment include [14], as discussed in section 2.4.

**2.2.2 Node Level.** Service deployment at the node level consists of installing and configuring the appropriate service logic on a node.

An in-band approach combines service logic and payload data in the same *active* packet. The service logic is executed in adequate execution environments (subject to availability) on each node the packet traverses.

An out-of-band approach retrieves the service logic from some code server or cache. Packets carrying payload data do not include the service logic. It is, however, possible that such packets carry references to service logic, which is subsequently installed on the node.

Some existing systems use active packets for in-band node-level deployment [8, 7, 13]. Other systems, such as [3, 6], apply active packets to out-of-band node-level deployment.

## 2.3  Distributed versus Centralized Service Deployment

The y-axis of the service deployment design space distinguishes between centralized and distributed service deployment mechanisms. In this context, the design choice refers to the method of deployment information processing.

### 2.3.1  Network Level.
Service deployment at the network level includes identifying suitable nodes that match service requirements.

A strictly centralized approach collects status and configuration information of nodes and topology information of the network at a central location, e.g. a network management station, and selects nodes that are to execute service logic.

In large networks, or when the network topology is highly dynamic — as is the case in ad-hoc networks — a distributed approach is required to cope with the complexity of the service deployment process. A distributed scheme may work similar to hierarchical routing algorithms. It extends the latter by more generic information exchange and aggregation, which includes not only link state information, but also node capabilities relevant to service deployment. Another approach to distributed network level service deployment may be based on mobile agents that cooperate in order to find a set of suitable nodes in the network. At the network level, active packets can also be classified as a distributed service deployment method because forwarding rules in active networks usually depend on node local decisions. An exception would be if some source-routing type of forwarding were used in the active packets.

### 2.3.2  Node Level.
Deploying services at the node level consists of selecting service logic that match node capabilities. Capabilities of nodes include the offered set of EEs and node-resident services, such as inter-EE communication facilities. At the node level, code modules must be installed and configured such as to correctly provide the specified functionality.

In a centralized approach this task is done by an entity similar to the network management station, which is also used for network level deployment. Another centralized approach to node level deployment is the case where a sender generates an active packet including code or direct references to code that can not be modified as the packet traverses the network.

In a distributed approach each node involved in a particular service performs the node level deployment separately. Such an approach may use a node independent service description, specifying the functionality to be implemented on a particular node. A distributed approach is also possible with active packets. That is, an active packet may modify its code or references to code based on some node state as it traverses the network. As a consequence the service logic that is finally deployed at a specific node depends on (distributed) state information.

At first sight, this dimension of the service deployment design space could also be considered as an early versus late binding issue. Using a node independent service description could also be classified as late binding, whereas including service code in active packets may be considered as early binding. In the case of active packets that are able to modify their own code, however, this classification would no longer be applicable. Therefore, we prefer the classify node level approaches into centralized and distributed.

## 2.4    Discussion of Existing Active Network Systems

In this section, we discuss existing active network systems from a service deployment perspective and classify them within the presented design space. The list of an analyzed systems is not exhaustive. The systems were chosen in a way to get a sampling of different design decisions. Table 1 summarizes our analysis, the details of which can be found in the following paragraphs. We observe that active networks typically use a distributed, in-band approach at the network level. Greater variety can be found in the approaches for the node level. The chosen approach depends on targeted services, as well as performance and security considerations.

| | network level | node level |
|---|---|---|
| ANN | in-band, distributed | out-of-band, either$^\alpha$ |
| ANTS/PAN | in-band, distributed | out-of-band, either$^\alpha$ |
| Chameleon | N/A | out-of-band, distributed |
| HIGCS | out-of-band, distributed | N/A |
| PLANet/SwitchWare | in-band, distributed | either$^\beta$, either$^\alpha$ |
| Smart Packets | in-band, distributed | in-band, centralized |
| Stream Code | in-band, distributed | in-band, either$^\alpha$ |

$\alpha$ : Choice between centralized and distributed is service specific
$\beta$ : Choice between in-band and out-of-band is service specific

**Table 1.** Classification of service deployment in existing systems

**2.4.1    ANN/Router Plugins.** The *Active Network Node (ANN)* [3] architecture makes use of active packets to deploy services. These packets feature a reference to a *router plugin*, which contains the service logic. If not cached locally on a node, router plugins are fetched from a code server and installed on the node.

Using active packets, the service deployment mechanism of this system can be classified as a *distributed, in-band* approach at the network level. It is distributed because the service logic is installed on active network nodes traversed by active packets. Whether an active network node is traversed or not depends on the forwarding tables in the nodes, which are set up by distributed routing algorithms. It is in-band because necessary information is contained in active packets, which also carry payload data.

At the node level, an out-of-band mechanism is used. That is, router plugins are fetched from a code server, using a different logical communication channel. As plugins may modify all fields of active packets, the choice between a centralized or distributed approach is left to the service designer, as explained in section 2.3.2.

**2.4.2    ANTS/PAN.** As far as service deployment is concerned, *ANTS* [6] and *PAN* [10] use a similar architecture. From a service deployment viewpoint, ANTS is similar to ANN. It also uses active packets, which contain a references to *code groups*, i.e. the service logic.

At the network level, the same comments as for ANN apply. At the node level, however, it is interesting to note that while using an out-of-band approach, ANTS efficiently mimics an in-band deployment mechanism. That is, if an active packet arrives at a node, service logic is — if not cached locally — retrieved from the cache of the previously visited node. Therefore, active packets and service logic generally follow the same path. The out-of-band approach, however, allows for an efficient use of network bandwidth because the service logic follows the first active packet of a stream. Subsequent active packets of the same stream will make use of the cached service logic. Therefore it is not necessary to transmit the service logic with each active packet. As in the case of ANN, the choice between a centralized or a distributed approach is left to the service designer.

**2.4.3    Chameleon.** *Chameleon* [11, 12] is a service deployment framework for the node level. The main goal of Chameleon is to support heterogeneous networks, that is a network may feature different flavors of active nodes. Therefore, it is not useful that active nodes get the service logic or a direct reference to it, because they are possibly lacking an adequate execution environment. Chameleon features a component-based service model. The *service creation engine* composes the service logic based on node independent service descriptions on each node separately. The components of the resulting service logic are dynamically loaded from code servers or caches. The number and type of components to be installed depends on node capabilities and types of available execution environments.

Chameleon does not describe the network level of service deployment. It may be combined with different existing approaches, e.g. a modified version of HIGCS (see 2.4.4) or using active packets containing a reference to a node independent service description.

At the node level, Chameleon represents a *distributed, out-of-band* approach to service deployment. It is obviously out-of-band as the service logic is dynamically loaded from code servers. More important is the fact that it is a distributed approach. Since each active node performs the node level deployment autonomously, service components that match node capabilities and take advantage of specific node features can be selected. The selection of components is constrained by node independent service descriptions.

**2.4.4    HIGCS.** In [14], the authors describe a network level service deployment framework for programmable networks. Although not targeted to active networks, a modified version of the main contribution of this work, the *Hierachical Iterative Gather-Compute-Scatter (HIGCS)* algorithm, may be applied to active networks as well. HIGCS allows to match node capabilities against service requirements, resulting in a set of appropriate nodes that are subsequently configured to implement the service.

HIGCS uses a *distributed, out-of-band* approach at the network level. Similar to hierarchical routing schemes, nodes build clusters and elect a cluster leader to aggregate information (e.g. node capabilities) and to represent it to the upper hierarchy level. The exchange is based on a specific control protocol (e.g. an extension to a hierarchical routing protocol) and therefore an out-of-band approach. It is distributed because cluster leaders process (aggregate, distribute) information relevant to service deployment within their cluster.

As HIGCS is intended for the network level deployment, a discussion of the node level service deployment is not applicable.

**2.4.5  PLANet/SwitchWare.** *SwitchWare* [8] is an architecture that combines active packets with *active extensions* to define networked services. Active packets contain code that may call functions provided by active extensions. Active extensions may be dynamically loaded onto the active node. *PLANet* implements this architecture using a safe, resource-bounded scripting language (called PLAN [4]) in the active packets and a more expressive language to implement active extensions.

At the network level, service deployment is implemented in a *distributed, in-band* way, using active packets similar to ANTS and ANN.

An interesting service deployment characteristic of the SwitchWare architecture is found at the node level. In fact, both in-band and out-of-band service deployment is used to combine the advantages of both worlds. Active packets contain code (in-band) that can be used like a glue to combine services offered by dynamically deployed active extensions (out-of-band). Similar to ANN and ANTS, the content of active packets may be modified by active extensions. Therefore, the choice between a centralized and distributed way of deployment is left to the service designer.

**2.4.6  Smart Packets.** *Smart Packets* [7] are active packets that contain code in a compact, machine-independent representation. *Sprocket*, the language Smart Packets are programmed in, is targeted to the management plane.

At the network level, Smart Packets use a *distributed, in-band* approach to service deployment. It is distributed because it makes exclusive use of standard IP forwarding at the network level. There is no central entity that determines on which nodes the service will be deployed (i.e. Smart Packets will be evaluated). It is in-band because the same mechanism (IP forwarding) is used for both packet forwarding and selection of nodes to run the service logic.

As opposed to ANTS and ANN, Smart Packets include the service logic in the active packet, not only a reference to it. As a consequence, they are a typical in-band service deployment representative at the node level. From an efficiency view point this makes sense because the management plane is targeted where no streams of packets are expected. That is, Smart Packet have the characteristics of agents that roam through the network and mainly interact with the Management Information Base (MIB) of the visited active nodes. The node level service deployment is organized in a centralized way: the originating node inserts the code to be executed on the active nodes in the active packet. Modification of this code is, to the best of our knowledge, not possible.

**2.4.7 StreamCode.** *StreamCode* [13] uses a hardware-decodable instruction set for describing the service logic in active packets. It has a StreamCode Execution Environment (SC-EE) where StreamCode programs are executed. This environment aims at high performance, and is basically designed for the data path. Since functions available in this environment are limited by program size restrictions and computational time bounds, complex services cannot be executed. Such services are provided in a EE, where long lived programs, such as routing daemons, are executed. EE programs may write their results into special area of the SC-EE. SC-EE programs may control themselves by reading out the result in the area.

At the network level, this system is a representant of a *distributed, in-band* approach to service deployment, with similar service deployment characteristics as other typical active packet based systems.

At the node level, we classify StreamCode as an *in-band* approach. It is in-band because the service logic is embedded in the packet. Active packets can not be directly modified, but they can be dropped and new packets may be generated by a service component in the EE. Hence, the choice between centralized and distributed deployment is left to service designer.

## 3  Integrated Service Deployment

In this section, we reason about the conceptual benefits of an integrated approach to service deployment. That is, why it is useful to use both in- and out-of-band service deployment in active networks. Furthermore, we present our approach to a hybrid system, which features an abstraction of service implementations, i.e. a service description language. We argue that the description language is necessary to cope with the requirements of real world scenarios.

### 3.1  Why Combining In- and Out-of-Band Service Deployment?

To better understand the benefits of a combined approach to service deployment, it is important to note that there are two equivalent models to program a distributed system, as observed by Wall in [2]. First, a *component based* model (sometimes also referred to as *discrete*) where service logic is installed on network nodes and exchange messages to be processed by service logic on other nodes. Second, an *active messages* based model where service logic moves from node to node and executes on the visited nodes. Wall observed that for a specific service (or distributed algorithm) one of those models is usually preferable over the other, because it leads to a more natural implementation of the algorithm. As a consequence, he suggested the development of a hybrid system that would naturally combine both programming models[1].

Being a generic platform for the implementation of distributed algorithms, active network nodes should — for the same reasons — support both programming models.

---

[1] In Wall's system, moving code exists as a concept only. In an implementation, he suggests to simulate moving code by pre-installing the code on all nodes and let the messages — similarly to ANN and ANTS — carry a reference to the code.

In fact, active networks can be considered as an architecture, which extends Wall's system — a combination of *active messages*[2] and component based logic — with dynamic service deployment. To support both programming models in active networks, a combination of in- and out-of-band service deployment is necessary.

On hybrid nodes, integrated service deployment may be performed in several ways. Firstly, embedded in the program, active packets may contain references to a service descriptor, which describes service logic to be installed out-of-band. That is, from a network level perspective, the service is deployed in-band. This deployment method is preferable if a maximum of flexibility for service definition is to be left to the end-user. Secondly, nodes are contacted by a management system requesting the deployment of service logic in a set of active nodes. In this scenario, network level deployment is performed out-of-band, which is advantageous if a network provider wants to keep full control over the network.

Architectures such as ANN, ANTS and SwitchWare, use a combination of in- and out-of-band service deployment, where, at the network level, in-band deployment is implemented with active packets and, at the node level, out-of-band deployment with some type of code components: (router plugins) in ANN, *code groups* in ANTS, and *active extensions* in SwitchWare. Having a closer look at those systems unveils some interesting aspects of combined approaches.

In ANN and ANTS, a combined approach is motivated by a gain in **efficiency**. More precisely, it enables code caching and, as a result, avoids latency for the installation of code and waste of bandwidth. That is, active packets contain references to code, which is retrieved from another cache or code server and cached locally. For subsequent packets containing the same references the locally available code can be used.

In SwitchWare, the main motivation is to allow for **secure extensibility** of the active packet programming language. That is, users are able to easily introduce new services in the network by programming active packets (in-band), whereas network-operators are enabled to extend the functionality of the packet programming language in a controlled, out-of-band way. Active packet programs are untrusted, but written in a restricted and safe language, while the components installed out-of-band are trusted, after having undergone some security checks.

In the next section, we propose an hybrid approach that combines in- and out-of-band deployment in a more generic way.

## 3.2   Integrated Service Deployment for Hybrid Systems

In this section, we argue that service deployment for hybrid systems has not been taken care of in a way that allows using them in real world scenarios. We illustrate an integrated approach to service deployment with our hybrid system based on Chameleon [12] and StreamCode [13].

---

[2] In the area of active networks, the term "active packets" is common to refer to a similar concept as Wall's "active messages". Henceforth, we will use active packets.

**3.2.1    The Chameleon/StreamCode Hybrid.** In our hybrid system *Chameleon* provides out-of-band service deployment, including a service description language dedicated to active networks. *StreamCode*, on the other hand, provides the means necessary for in-band deployment, as discussed in section 2.4.

The Chameleon/StreamCode hybrid node (cf. Figure 2) is characterized by at least two EEs: an active packet EE and a component-based EE.

EEs are usually targeted to a certain type of functionality. As shown in section 2.4, some EEs are optimized for the management plane, whereas others are adapted for the control or even data plane. There is generally a tradeoff between the abstraction level (or richness) of the API and the performance provided by an EE. Since the requirements for an EE vary widely from plane to plane, there is no single EE design that is superior to others for all interesting services. As a consequence, our active node must provide more than one EE, if it is to support the whole spectrum of services.

In our active node implementation, the StreamCode-EE (SC-EE) plays the role of an active packet EE. The component based EE uses Java as the underlying technology. Those two EEs are required for the separation of data and control/management plane. Communication between EEs is possible through a reserved area of shared memory. In particular, StreamCode programs may access results of complex computations of service logic in the component-based EE. Furthermore, the component based EE may generate StreamCode packets to communicate with the SC-EE. In this way the computations in the data plane are decoupled from the processing in the control/management plane.

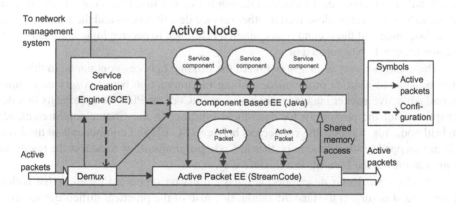

**Fig. 2.** Functional architecture of a Chameleon/StreamCode hybrid node.

Up to this point, our node is, from an architectural viewpoint, similar to ANN or SwitchWare nodes. There are, however, two distinctive architectural differences.

Firstly, Chameleon provides means to abstract service functionality from a service implementation. This abstraction comes in the form of a service description language.

Secondly, Chameleon/StreamCode nodes feature a *service creation engine* (SCE). The SCE is able to interpret *node independent* service descriptors and, based on this, to compose a service implementation that is *dependent* on the node environment.

An important consequence of the abstraction provided by the service descriptors and the node local mapping process (provided by the SCE) is the ability to deal with heterogeneous active nodes. That is, with active nodes running sets of EEs that may vary from node to node. Another benefit of the proposed architecture is the ability to compose services from components, which, in turn, enables code reuse.

We describe these features and their benefits in the following paragraphs.

**3.2.2   Active Network Service Description Language (ANSDL).** Particular service logic may be implemented in different ways, depending basically on the set of EEs available on a specific node. The role of the ANSDL is to abstract service functionality from its implementation. The main goal is to provide a description of service logic in a node independent way.

In [12], we proposed an XML-based language to address this issue. The details of the language are beyond the scope of this paper. We provide, however, a short overview.

The ANSDL centres around two main abstractions: *containers* and *connectors*. Containers abstract service logic as a black box with interfaces. Connectors abstract the binding among containers. Containers can be composed of other (sub-)containers bound to each other via connectors. We assume that container names capture the functionality of the service logic.

There are two types of service description documents, both include a container name and number and types of interfaces. Documents of the first type contain information about sub-containers — described in other service descriptors — and their interconnection. Documents of the second type contain a reference to service logic and the type of EE the logic is to be executed in.

The connectors act as an abstraction of the binding between containers. In this way, it is possible to abstract protocols describing the interaction between service components. An active packet language, such as StreamCode or Plan, complements this description with the possibility for dynamical binding. In the Chameleon/StreamCode hybrid node, for example, the connectors between SC-EE and components in the Java-EE are mapped to shared memory. StreamCode programs allow to access the results of computations in the Java-EE in a flexible way.

Clearly, a particular description language must be supported by all active nodes. Therefore, it is subject to standardization. Because of the practical difficulties to agree on a standard node environment, we believe it is much easier and more reasonable to standardize a description language.

**3.2.3   Service Creation Engine (SCE).** Each active node is responsible for selecting suitable service logic by matching the container's requirements (attributes in the descriptors) against node capabilities. During the matching process, descriptors of sub-containers are interpreted until all descriptors contain direct references to service logic (e.g. router plugins, code groups, active packet code etc.). This mapping task is performed by the SCE, which is available on each node. The result depends on the node environment (available EEs, node-resident services, etc.)

## 4  Conclusion

In this paper, we explored the design space of service deployment in active networks and proposed a new scheme to classify different approaches according to their 1) service distribution mechanism and their 2) method of deployment information processing. The proposed scheme was then applied to a number of service deployment approaches described in the literature. As to our knowledge, there is no previous work that explores the design space of service deployment in active networks in a systematic way.

Moreover, we presented an approach showing how active packet based networks can be integrated with an out-of-band service deployment framework. We argued that both active packet based networks — which are a typical realization of in-band service deployment — and component based technology, using out-of-band service deployment have advantages and shortcomings. Fortunately, they are rather complementary. Hence, integration is beneficial as it results in a highly flexible, though efficient network architecture. Furthermore, we described our own hybrid approach to service deployment that can deal with heterogeneous active networks, because the service deployment framework is able to interpret abstract, i.e. node independent, service descriptions.

## 5  Acknowledgments

We would like to thank Michael Hicks and the anonymous reviewers for the detailed and insightful comments on an earlier version of this paper.

## References

[1] D. Wetherall. Active network vision and reality: lessons from a capsule-based system. In Symposium on Operating System Principles (SOSP'99), December 1999.

[2] D.W. Wall. Messages as Active Agents. In ACM Symposium on Principles of Programming Languages (POPL), Albuquerque, New Mexico, January 1982.

[3] Dan Decasper, Guru Parulkar, Choi, S., DeHart, J., Wolf, T., Plattner, B., A Scalable, High Performance Active Network Node, IEEE Network, Vol. 13(1), 1999.

[4] M. Hicks, P. Kakkar, J.T. Moore, C.A. Gunter and S. Nettles. PLAN: A Packet Language for Active Networks. In ACM SIGPLAN International Conference on Functional Programming Languages, 1998.

[5] M. Hicks, J.T. Moore, D.S. Alexander, C.A. Gunter, S.M. Nettles. PLANet: An Active Internetwork. In IEEE Infocom Conference, 1999, New York, USA.

[6] D. J. Wetherall, J. V. Guttag, D. L. Tennenhouse. ANTS: A Toolkit for Building and Dynamically Deploying Network Protocols. In IEEE Openarch'98, San Francisco, USA, April 1998.

[7] B. Schwartz, A.W. Jackson, W.T. Strayer, W. Zhou, R.D. Rockwell, C. Partridge. Smart Packets for Active Networks. ACM Transactions on Computer Systems, Vol. 18(1), February 2000.

[8]  D.S. Alexander, W.A. Arbaugh, M.W. Hicks, P. Kakkar, A.D. Keromytis, J.T. Moore, C.A. Gunter, S.M. Nettles, J. M. Smith. The SwitchWare Active Network Architecture. IEEE Network Special Issue on Active and Controllable Networks, vol. 12 no. 3, pp. 29 - 36.

[9]  D. Decasper, Z. Dittia, G. Parulkar, B. Plattner. Router Plugins - A Software Architecture for Next Generation Routers. IEEE/ACM Transactions on Networking, February 2000.

[10]  E.L. Nygren, S.J. Garland, M.F. Kaashoek. PAN: A High-Performance Active Network Node Supporting Multiple Mobile Code Systems. In IEEE Openarch'99, March 1999.

[11]  M. Bossardt, L. Ruf, R. Stadler, B. Plattner. Service Deployment on High Performance Active Network Nodes. In IEEE NOMS 2002, Florence, Italy, April 2002.

[12]  M. Bossardt, L. Ruf, R. Stadler, B. Plattner: A Service Deployment Architecture for Heterogeneous Active Network Nodes. In 7th IFIP SmartNet 2002, Saariselkä, Finland, April 2002.

[13]  T. Egawa, K. Hino and Y. Hasegawa. Fast and Secure Packet Processing Environment for Per-Packet QoS Customization. In IWAN 2001, September 2001.

[14]  R. Haas, P. Droz, B. Stiller. Distributed Service Deployment over Programmable Networks. In DSOM 2001, Nancy, France, 2001.

[15]  D.L. Tennenhouse, and D.J. Wetherall. Towards an Active Network Architecture. Computer Communication Review, Vol. 26, No. 2, April 1996.

# Component-Based Deployment and Management of Services in Active Networks

Marcin Solarski[1], Matthias Bossardt[2], and Thomas Becker[1]

[1] Fraunhofer Institute for Open Communication Systems FOKUS
Berlin, Germany
{solarski,becker}@fokus.fhg.de
[2] Computer Engineering and Networks Laboratory
Swiss Federal Institute of Technology, ETH
Zürich, Switzerland
bossardt@tik.ee.ethz.ch

**Abstract** This paper [1] describes a holistic approach towards the deployment and runtime management of services on active network nodes taken by the FAIN project. Both the underlying service model and the architectures supporting deployment and management are component oriented. The separation of service meta-information and implementation code allows for a very flexible way of service deployment management as it facilitates selective code distribution, fine-grained installation and instantiation. Active services are composed from a set of service components that can be selected on demand at deployment time and installed in any combination of the data, control, and management planes which enables realisation of arbitrary active services.

## 1 Introduction

Since the mid nineteen-nineties many efforts have been made to develop Active Networks technology [1] to enable more flexibility in provisioning services in networks. By defining an open environment on network nodes this technology allows to rapidly deploy new services which otherwise may need a long time and adaption of hardware.

There are two major approaches to Active Networks: on the one hand active packets (*capsules*) transmitting code along the data path which will be executed on the appropriate nodes and on the other hand a separated transmission of control and data packets. Whereas the first approach is suitable for deploying simple data path services (like new forwarding rules) with low latency, the latter one is more applicable for high-level, application-oriented services.

Deploying and managing high-level services requires an appropriate service model. While fully-fledged component-based service models are an integral part of many enterprise computing architectures (e.g. Enterprise JAVA Beans, CORBA Component Model, Microsoft's .NET), it is not the case in many approaches developed by the active networking community.

---

[1] This research is funded by the European Commission (IST-1999-10561), as well as by the Swiss Federal Institute of Technology (ETH) Zürich, and Swiss BBW (grant number 99.0533).

J. Sterbenz et al. (Eds.): IWAN 2002, LNCS 2546, pp. 87–98, 2002.
© Springer-Verlag Berlin Heidelberg 2002

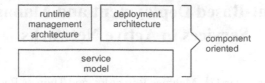

**Fig. 1.** Component oriented service model and supporting architectures

In this paper we present a *component-based* service model together with deployment and runtime management architectures based on this model (see figure 1). While the service model defines how services are described by potential recursive compositions of service components the deployment architecture defines how and when service components are transfered to active nodes and installed upon a service request. The runtime management architecture deals with the installation of serivce components in execution environments and the management of component instances.

The mentioned architectures have been implemented during the still ongoing FAIN project and used in the project test bed.

Section 2 describes the service model in more detail. Sections 3 presents the design and implementation of the runtime management architecture while section 4 presents the deployment architecture together with an example. In section 5 we show some related work and in section 6 we draw a conclusion and give an outlook on future work.

## 2   Service Model

This section describes the FAIN service model. The basic concept of the service model is the service *component*, a piece of self-contained software that is the smallest unit of deployment and management. A *service* is a unit of functionality that a service provider wants to offer to its customers. This functionality is realised by a combination of one or more service components.

The composition pattern is hierarchical in that service components may be recursively composed of sub-components. The goal of this approach is to enable flexible deployment and management of services at a suitable level of granularity, as well as to deal with services running in multiple execution environments with varied capabilities and underlying software technologies on possibly heterogeneous active nodes.

From the **deployment perspective** , a service component consists of a service descriptor and an optional reference to the service code stored in a file called *code module*. There are three classes of service components that differ in the terms of whether they consist of some service subcomponents and whether they directly refer to a code module: *(Simple) Implementation* – a service component without any dependencies. It contains just a reference to a code module; *Compound Implementation* – a service component consisting of subcomponents and having a reference to a code module; and *Abstract Implementation* – a service component consisting of subcomponents and having no reference to a code module.

From the **runtime perspective** a component is a unit of instantiation and activation. Runtime instances of components are identified by a unique name and are owned by an identity defined by its name and credentials. Assigning an owner to each component instance allows to do access control and accounting based on identities. The configuration of a component instance is defined by a set of properties. The connections to the outside are described by the component's ports where components can offer arbitrary ports in addition to the basic ones, e.g. a bandwidth manager will offer an interface for reserving bandwidth in addition to the initial interface.

Components are accessed and interconnected by *ports*. Ports can be used for exchanging information or to model this exchange. Ports are also useful to express dependencies among components. A port has a name valid in the context of the holding component as well as a reference to the component itself. A port is described by a direction, an address (i.e. the endpoint for data exchange like an IP address, a memory address, an IOR, etc.), a format (i.e. the protocol used for data exchange like IP, ATM, IIOP, HTTP, etc.), and an optional type used for typed ports (e.g. a CORBA interface repository ID). The term "port" is used here to generalise from the notion of "interface" used in distributed object technology.

Each component offers a special port, namely its *iComponentInitial* interface. This interface is used for the initial access of a client to a component. Using this interface a client can retrieve the references to the other ports (and interfaces) supported by the component. In order to get access to a port of the component the client has to authenticate itself at this interface passing its credentials. The result is a reference to the requested specific interface tailored to the calling client. Components can offer arbitrary ports in addition to the basic ones, e.g. a bandwidth manager will offer an interface for reserving bandwidth in addition to the initial interface.

The management of components has two aspects. The component life cycle is managed using *component managers* as it is described in section 3. The other aspect is called *component configuration* and is a process of setting up or tuning the component behavior that may take place after the component is instantiated. The configuration of a component is described by properties as pairs of names and values. Properties can be used to define a component's behavior, e.g. a property may define a resource limit for a particular user or may restrict the access and usage of a component's interface. Interested clients can register for getting notified when particular properties change.

## 3   Node Management Architecture

This section describes the design and implementation of the runtime management architecture as it was developed during the FAIN project [13]. Since this architecture is based on the component-orineted service model its basic abstraction is as well the notion of components. Before presenting the desing in more detail a short introduction to the runtime environment for active services will be given.

The places where service code is executed are called *execution environments*. While this is common to all active network approaches the FAIN runtime architecture defines additionally the notion of *virtual environments*. Virtual environments (VEs) were introduced to abstract from the specifica of execution environments (EEs), for example, there are EEs implemented in hardware offering high performance, other EEs are implemented in interpreted languages focusing on flexibility.

A VE is owned by a service provider and may group several EEs assigned to the particular service provider. Thus, a VE is the "frontdoor" to the services running in the execution environment(s) of a service provider and used for their management. There is one special *privileged* VE owned by the node provider which holds the basic node services, like EE management, VE management, bandwidth management, etc. The priviledged is the main entry poitn for an active node.

When a service provider owns multiple VEs spread over several network nodes the VEs form a virtual active network. In order to identify which VEs belong to a virtual network they are tagged with a unique virtual network identifier.

## 3.1 Design

To support the component-based service model the design of the runtime management architecture also uses the component as the main abstraction. From the management viewpoint a component instance represents two aspects: firstly the functional aspect where the component is seen as a service instance and secondly the non-functional aspect where the component is seen as a resource.

The management architecture utilizes a special kind of component ports, namely *interfaces* (as known from CORBA [10]), and defines a basic set of interfaces which comprise the management API of the active node. They are used for example by the active service provisioning (ASP) module (see section 4.1) to install services on the node and later to create instances. Service instances use the same API to discover other services or resources. Because of the reflective nature of this API service instances can retrieve and modify metainformation about themselves.

The three basic types of interfaces used for the runtime management are *iTemplateManager*, *iComponentManager*, and *iComponentInitial*. The runtime management allows adding new ports to components so that they can publish their specific functionality. This allows for a very flexible way to construct services from basic components by combining and enhancing the already provided functionality.

In the following the abstractions defined by the runtime anagement will be described in more detail.

**Template Manager.** A template manager manages templates (e.g. JAVA classes, object files, etc.) from which component instances can be created. Management comprises the installation, removal, and updating of templates. This operations are available at the template manager's *iTemplateManager* interface. Template managers are implemented by VEs and EEs as the environments are the places where service components are installed. The active service provisioning (ASP) module will use the template manager of the priviledged VE to install a new template on the node and pass a description of the template.

A template description includes all the information which is necessary to create component instances. However, instead of having the template manager to deal with different instantiation methods the template description contains a factory for instances, called component manager. Further the template description includes the name and version of the template, the VE and EE identifiers, the path to the code archive of the template, and optional additional properties defining template specific features. The VE and EE identifiers are used to determine the service provider for whom to install the template and the runtime environment for component instances (e.g. a JAVA virtual machine).

**Component Manager.** A component manager is used to manage instances of components associated to a specific template, thus it is acting as a component factory. Managing comprises the creation, activation, deactivation, discovery, deletion, and updating of instances. This operations are available at the component manager's *iComponentManager* interface.

The parameters for the creation of a component are specific to the type managed by the component manager (as defined by the respective template) and the result is the component's *iComponentInitial* interface. The component manager uses the parameters to check the possibility to create a new instance, i.e. the availability of the required resources.

**Specific Component Managers.** While template managers are only specific to the EE in which templates should be eventually installed the component managers are even more specific. For each type of components there needs to be a component manager. When a particular type of resource is represented by a type of component (e.g. a process represented by a component) there has to be the appropriate resource manager (e.g. a process manager). In order to define a manager one should specify the additional interfaces and operations if any, the properties supported for resource creation, and the dimensions and units supported for monitoring resources if applicable.

There are special managers for managing the basic services provided by an active node, namely security management, traffic management, packet demultiplexing, and management of execution environments as well as virtual environments.

**Example.** Figure 2 depicts an example snapshot of an active node. From left to right it shows the priviledged VE with the attached priviledged EE supporting *iTemplateManager* (iTM) interfaces. Inside the priviledged EE there are installed three templates with respective component managers supporting *iComponentManager* (iCM) interfaces: VE management, EE management, and management of an arbitrary resource "z". The VE manager created one VE, the EE manager created two EEs, and the arbitrary resource manager one resource instance. Both EEs and the resource instance are attached to the VE. Further, it can be seen that in EE1 there is installed a template "x" and in EE2 a template "y". From both templates an instance was created.

## 3.2   Implementation

To support the management of services and resources on an active node a framework was implemented comprising base classes for components, component managers, and template managers. The implementation was carried out in JAVA with a strong support for CORBA interfaces as communication ports (although components can implement

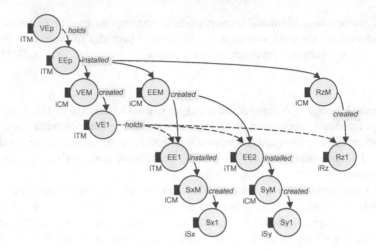

**Fig. 2.** Example snapshot of components on active node (iTM = iTemplateManager, iCM = iComponentManager, iSx = interface of service "x", iSy = interface of service "y", iRz = interface of resource "z")

whatever type of port is most appropriate). This framework facilitates the implementation of JAVA based services as well as services based on different technology by creating wrapping components.

Using the framework there were implemented a collection of basic services to support other services which would be deployed to the node dynamically. Those basic services live in the node provider's privileged execution environment and are thus able to access operating system resources. Other services living in the service providers' environments can use the functionality provided by the basic services after successfully accessing their interfaces.

## 4   Service Deployment Architecture

Service deployment is considered in this paper a process of making a service available in the active network to the service user. It involves determining the target environment, identifying the service components needed, mapping them to the target environment, fetching the code, as well as installing and activating service components in their target environments.

This section describes the FAIN service deployment architecture called Active Service Provisioning (ASP) system. The system is defined by its main functionalities in section 4.1. The design of the ASP system, including the architectural components and their interrelationship is the contents of section 4.2. Finally, the operation of the ASP system is explained in a concrete deployment scenario for an example transcoder service in section 4.3.

## 4.1   Functionality of the Active Service Provisioning System

In this section, the main functionality of the Active Service Provisioning system is described. The main actors communicating with the ASP system are:

**Service Provider,** or SP for short, composes services that include active components and deploys these components in the network via the Active Service Provisioning, and offers the resulting service to Consumers.

**Active Network Service Provider,** or ANSP for short, provides facilities for the deployment and operation of the active components into the network. Such facilities come in the form of an active middleware, support of new technologies etc.

The ANSP owns an active network offering one or more environments where active code from Service Providers can run.

These roles are described in the FAIN Enterprise Model in [12] in more detail. The main use cases of the ASP system are:

*Releasing a service.* The Service Provider who decides to offer his service in the active network has to release it in the active network. The service is released by making the service meta-information and service code modules available to the ASP system.

*Deploying a service.* After the service is released in the network, the Service Provider may want to deploy his service so that it can be used by a given service user. It means finding a target environment that is most suitable for the given user, determining a mapping of the service components to the available EEs of the target environment, downloading the appropriate code modules, and finally installing and activating them.

*Removing a service.* The Service Provider may request to remove a deployed service from the environment it was deployed in. The ASP identifies the installed service components and removes them from the EEs of the target environment.

*Withdrawing a service.* A service released in the active network may be withdrawn so that is is not available to be deployed any more. The ASP removes the service metainformation and discards the service code modules.

## 4.2   Design

The design of the ASP system follows a two tier approach. We distinguish between network and node level ASP system. The network level functionality consists of metainformation and code module management, as well as the selection of nodes that are to execute part of the service logic. On the node level, necessary service components are identified, and dependencies are resolved. What follows is a more detailed description of the ASP design, and how the functionality presented in section 4.1 is achieved by our implementation.

**Service Descriptor.**   The service descriptor supports a service model as described in section 2. The first part of the service descriptor holds information about the developer or provider and the functionality of the corresponding service component. The second part is dependent on the class (cf. section 2) of service component described. For a *simple implementation* this part contains a reference to a code module and identifies the target environment where the code module is to be installed. It further contains EE-specific information, which is used to perform EE-specific part of deployment process. The service descriptor of an *abstract implementation* holds information about required subcomponents and how they are to be bound to each other in order to perform the expected functionality. Finally, a *compound implementation* is a mixture of the two classes above, and hence contains both sets of information. The service descriptor is implemented in XML, which proved to be a very suitable technology for the task at hand. Furthermore, we developed an XML Schema to verify the structure and correct syntax of service descriptors.

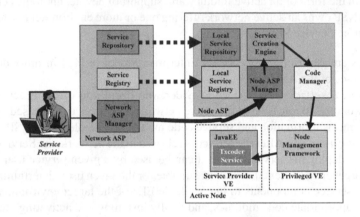

**Fig. 3.** ASP interaction when deploying an example service

**Network Level ASP Design.**   The network level ASP system consists of three components depicted in figure 3: Network ASP manager, Service Registry and Service Repository.

The **Network ASP Manager** serves as an access component to the ASP system. In order to initiate the deployment of an particular service, a Service Provider contacts the Network ASP Manager and requests a service to be deployed as specified by the service descriptor.

The **Service Registry** is used to manage service descriptors. Service descriptors are stored on it, when a service component is released in the network. Network ASP Manager and the Service Creation Engine may contact the Service Registry to fetch service descriptors.

The **Service Repository** is a server for code modules. A code module is stored on the Service Repository, when a service descriptor referencing the particular code module is released in the network. The Code Manager, which is part of the node level ASP system, may fetch code modules from the Service Repository.

**Node Level ASP Design.**   On the node level, the following components make up the ASP system as shown in the *node ASP* block of figure 3: Node ASP manager, Service creation engine and Code Manager.

The **Node ASP Manager** is the peer component to the network ASP manager on the node level. The network ASP manager communicates with the node ASP manager in order to request the deployment, upgrading and removal of service components.

The **Service Creation Engine** (SCE) selects appropriate code modules to be installed on the node in order to perform the requested service functionality. The service creation engine matches service component requirements against node capabilities and performs the necessary dependency resolution. More details about this mapping process can be found in [8], [9]. Since the service creation engine is implemented on each active node, active node manufacturers are enabled to optimize the mapping process for their particular node. In this way it is possible to exploit proprietary, advanced features of an active node. The selection of service components is based on service descriptors.

The **Code Manager** performs the execution environment independent part of service component management. During the deployment phase, it fetches code modules identified by the service tree from the service repository. It also communicates with Node Management to perform EE-specific part of installation and instantiation of code modules. The Code Manager maintains a database containing information about installed code modules and their association with service components.

## 4.3   Deploying an Example Service

This section describes a scenario in which an example active service is being deployed. After the structure of the service is presented, the details of the ASP components interactions are given.

**Example Service.** In order to evaluate our architecture, a transcoder service has been implemented [15] and deployed in the FAIN test bed (cf. figure 5). The service functionality is implemented in two code modules, a transcoder controller module and a transcoder engine module. Both are to be deployed on the same node. Three service descriptors are used to hold the corresponding meta-information. The first service descriptor holds information about an *abstract implementation* of a *transcoder*. In particular, it contains references to two sub-services – a transcoder controller and a transcoder engine – which make up the final transcoder service. Both sub-services, transcoder controller and transcoder engine, may be *abstract implementations* themselves. In the transcoder version we implemented, however, the sub-services are *simple implementations*. That is, each sub-service consists of service descriptor and associated code module. The service descriptors for *simple implementations* (cf. figure 4.3) holds both EE-unspecific and EE-specific information. The EE-unspecific part identifies, among others, the EE in which a particular code module is supposed to be executed, as well as the name and/or location of the code module. For our Java-based EE, the EE-specific part, which can be found in the *PROPERTIES* section, contains information such as codepath, and main class name, and others.

Figure 5 presents the configuration of the FAIN test bed and the system components involved in the deployment process. The FAIN test bed contains active nodes located at FhG, ETH and UPC, the FAIN Consortium partners' sites. On each of the active nodes, both the Node Management Framework and node-level Active Service Provisioning

```
<?xml version="1.0" encoding="UTF-8"?>
<SERVICE xmlns:xsi="http://www.w3.org/2001/XMLSchema-instance"
xsi:noNamespaceSchemaLocation="C:chameleon.xsd" xsi:type="IMPLEMENTATION">
    <DESCRIPTION>
        <SERVICE_NAME>transcoder_controller</SERVICE_NAME>
        <SERVICE_ID/>
        <PROVIDER>FAIN</PROVIDER>
        <VERSION>0.1</VERSION>
    </DESCRIPTION>
    <PROPERTIES>
        <PROPERTY>
            <KEY>mainClassName</KEY>
            <VALUE>org.ist_fain.services.transcoder_controller.
            TranscoderManager</VALUE>
        </PROPERTY>
        <PROPERTY>
            <KEY>mainCodePath</KEY>
            <VALUE>/usr/local/jmf-2.1.1/lib/jmf.jar:/usr/local/jmf-2.1.1/lib/
            sound.jar:/usr/local/jmf-2.1.1/lib</VALUE>
        </PROPERTY>
    </PROPERTIES>
    <ENVIRONMENT>
        <EE_NAME>JVM</EE_NAME>
        <EE_VERSION>1.3.1</EE_VERSION>
    </ENVIRONMENT>
    <CODE xsi:type="CODE_LOCATION">
        <CODEBASE>jvm.transcoder1.FAIN.transcoder1.jar</CODEBASE>
    </CODE>
</SERVICE>
```

**Fig. 4.** XML service descriptor example

systems are running. The network ASP components are located so that the Service Repository is at UCL in London, the Service Registry at FhG and the Network ASP Manager at UPC.

In an example scenario, the Service Provider requests deployment of his transcoder service. The service is composed of two service components that need to run colocated on a Java-enabled active node. The network ASP decides that the optimal target environment is the active node at ETH and sends a node-level deployment request to the node ASP on the target node. The node ASP processes the requests by resolving service dependencies, fetching the needed code modules and triggering the EE-specific installation process. Finally, a transcoder service instance is created.

Now the service is ready to use. The service provider may configure the transcoder to convert a JPEG video stream into a H.263 format and to forward the video data to a given receiver.

## 5   Related Work

The work of Marcus Brunner et al. on **Virtual Active Networks** [2] defines an architecture for the creation and management of services in an active network. However, it lacks the component-oriented service model allowing to flexibly combine functionality out of service components. The framework defined by the **IETF ForCES** [4] separates functionality found on a network node into forwarding and controlling elements. Further,

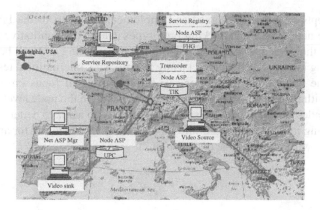

**Fig. 5.** Deploying the transcoder service – component distribution

ForCES aims at defining a model which describes how those elements are connected to form a self-contained network element.In comparison to our work ForCES seems to have a narrower scope focussing on the control and forwarding planes. **NetScript** [6] uses a recursive mechanism for the composition of services from components. However, it does not deal with the service deployment process and does not support multiple, concurrent execution environments. The **LARA active router architecture** developed at the University of Lancaster [5] is also based on components. This architecture allows to create services from components in a flexible and extensible way. However, this work focuses on the runtime aspects and doesn't define a mechanism to map a service description to interconnected component instances. The **AMNet project** now continued in the flexinet project [7] defines an architecture for programmable networks. Service modules can be loaded from a repository to active nodes where they run inside execution environments on top of a resource control layer. Although it is possible to combine modules out of multiple object code pieces there seems to be no explicit component model for services. The **CORBA Component Model** (CCM) [11] defines languages to specify components and their implementations. Further it defines a deployment and runtime environment for components. On the other hand it concentrates on enterprise applications and sticks to a client-server model. It doesn't allow to introduce new kinds of ports (e.g. stream oriented) for connecting components.

## 6  Discussion and Future Work

The approach presented in this paper covers the whole process of handling services in active networks beginning with a service model and stretching to deployment and runtime architectures. The choice of a component-based approach facilitates the fine-grained service description, deployment, and management. In the completeness of our approach, not focussing on one aspect while neglecting others, we see its novelty. Particular features are: out-of-band deployment, separation of service metadata and code, node-level service component dependency resolution, component-oriented service model and support for heterogeneous service implementations.

Our concepts have been designed and implemented as part of the FAIN architecture for active networks. First working prototypes can give a qualitative proof for the feasablitity of the design. A quantitative evaluation is pending and will be tackled during the ongoing last year of the project using the project's international testbed.

Besides the quantitative evaluation our future work will focus on extending our architecture with regard to its flexibility and robustness. On the node level, we intend to consider upgrades of running active services on the fly to achieve better availability of rapidly changing implementations of the services. Further, we target for an optimisaton of the network-wide service code distribution.

# References

1. Tennenhouse, D.L., Wetherall, D.J.: Towards an Active Network Architecture. Computer Communication Review, Vol. 26, No. 2, April 1996
2. Brunner, M., Plattner, B., Stadler, R.: Service Creation and Management in Active Telecom Networks. Communications of the ACM, April 2001
3. Galis, A., Plattner, B., Moeller, E., Laarhuis, J., Denazis, S., Guo, H., Klein, C., Serrat, J., Karetsos, G., Todd, C.: A Flexible IP Active Networks Architecture. International Working Conference on Active Networks (IWAN 2000), Tokyo, Japan, October 2000
4. IETF ForCES Working Group: Forwarding and Control Element Separation, http://www.ietf.org/html.charters/forces-charter.html
5. Schmid, S., Finney, J., Scott, A.C., Shepherd, W.D.: Component-based Active Network Architecture. Proceedings of 6th IEEE Symposium on Computers and Communications, 3-5 July 2001
6. Da Silva, S., Florissi, D., Yemini, Y.: Composing Active Services in NetScript. Position paper, DARPA Active Networks Workshop, Tucson, AZ, March 9-10, 1998.
7. Fuhrmann, T., Harbaum, T., Schller, M., Zitterbart, M., AMnet 2.0: An Improved Architecture for Programmable Networks, submitted to IWAN'02, available from www.flexinet.de
8. Bossardt, M., Ruf, L., Stadler, R., Plattner, B.: A Service Deployment Architecture for Heterogeneous Active Network Nodes. Kluwer Academic Publishers, 7th Conference on Intelligence in Networks (IFIP SmartNet 2002), Saariselkä, Finland, April 2002
9. Bossardt, M., Ruf, L., Stadler, R., Plattner, B.: Service Deployment on High Performance Active Network Nodes. IEEE Network Operations and Management Symposium (NOMS 2002), Florence, Italy, April 2002
10. OMG: The Common Object Request Broker: Architecture and Specification. http://www.omg.org/cgi-bin/doc?formal/02-05-15.pdf, Mai 2002
11. OMG: CORBA Components final submission. http://www.omg.org/cgi-bin/doc?orbos/99-02-05
12. FAIN Deliverable D1: Requirements Analysis and Overall Architecture. FAIN Consortium, May 2001, pp. 11-18
13. FAIN Deliverable D4: Revised Active Node Architecture and Design. FAIN Consortium, May 2002
14. FAIN Deliverable D5: Revised Specification of Case Study Systems. FAIN Consortium, May 2002
15. FAIN Deliverable D6: Definition of Evaluation Criteria and Plan for the Trial. FAIN Consortium, December 2001

# ANQL – An Active Networks Query Language

Craig Milo Rogers

USC Information Sciences Institute
4676 Admiralty Way
Marina del Rey, CA, USA
rogers@isi.edu

**Abstract.** This paper discusses parallels between network communication packets, when processed in bulk, and relational database records. It introduces a new application-specific language, ANQL (Active Networks Query Language), that exploits a database metaphor for packet processing. ANQL has been demonstrated in Active Network control and management plane activities, although it may also be used in many other networking applications. In active networks, ANQL is primarilly intended as a tool or adjunct for use by Active Applications, and by control and management code. Environments are discussed in which ANQL or related languages might be utilized as full-fledged active packet languages in themselves. ANQL is applicable to both event-driven and background processing activities, and may be used in a single, centralized data collection and analysis process, or, with little change, in distributed implementations of packet analysis activities.

## 1 Introduction

There are many tasks that need to be performed in the control and management planes of a data communication network. Some tasks include extracting or acting on packets that meet certain conditions; representative tasks include:

- checking for faults
- detecting intrusions
- monitoring content for viruses
- identifying packets for which special routing or services are applied

Other tasks may require taking actions based on aggregate values calculated on groups of packets. These tasks may include:

- sending an alarm when a sequence of packets indicates that an attack is in progress
- detecting when excessive traffic of a particular class is monopolizing network resources
- reporting bandwidth utilization statistics

J. Sterbenz et al. (Eds.): IWAN 2002, LNCS 2546, pp. 99–110, 2002.

Problems of this sort have been present since computer networks first became available, and many solutions have been implemented in the past. The advent of Active Networking [14] [12] gives us a new set of tools that we may apply to the control and management of computer networks, providing opportunities to create better solutions to long-existing network management problems.

## 2   Comparing Packets to RDBMS Data

The central metaphor in this paper may be summarized as follows:

> Computer network packets are analogous to table joins in a
> relational database management system (RDBMS).

In the relational database model [5], data records are grouped into tables of records with a similar structure. Data records are extracted from one or more tables by selecting records with specific values in certain fields. Records from different tables are *joined* by matching records according to the values of specific fields. The results from a table join can be utilized on a record-by-record basis, or condensed into summary information.

Although it may not be apparent at first glance, collections of network communication packets are very similar to relational database data. Consider a set of packets $P$ constructed from a fixed protocol stack, such as **ipv4/udp/anep**. Each packet in $P$ consists of one instance each of an IPv4 header, a UDP header, and an ANEP header, stored adjacently in the packet. Each of the protocol headers consists of a set of fields. The set of IPv4 headers for all packets in $P$ is analogous to the records of an **ipv4** relational database table, using the same set of fields; the set of UDP headers are analogous to the records of a **udp** table, and similarly for the set of ANEP headers.

Let us define each of the headers to have a virtual field, **packet_id**, containing a value that is unique to each packet in $P$; all headers in a single packet will have the same value for **packet_id**. Under this definition, the set of **ipv4/udp/anep** packets in $P$ may be said in RDBMS terms to be a *prejoined* (also called *clustered*) physical storage representation for the logical records of the separate **ipv4**, **udp**, and **anep** tables, with **packet_id** as the *cluster key*.

Of course, there are differences in usage between the conventional relational database model and collections of packets. RDBMS systems usually operate on data that has been indexed for efficient access, while network packets are commonly acquired in real-time event streams (although it is not uncommon to collect packet traces and perform retrospective analyses). Many network protocols, such as IPv4, TCP, and ANEP, contain varying-length optional contents, the structure of which might best be *normalized* into multiple tables in an RDBMS model. However, as will be shown, it is possible to gain a considerable advantage from the relational database model of network packets without addressing these concerns in detail.

This completes the mapping between network packets and RDBMS records as stored in a typical RDBMS. By recasting network packets into the relational database model, we gain access to the tools and methodologies that have been developed in the last three decades for processing data in this form. In particular, we have access to the database query language SQL (Structured Query Language) [13] and tools based on it.

# 3   Applying the RDBMS Model

The Active Networks Query Language (ANQL) is an SQL-like application language that can be used to extract, summarize, and reformat information about packets, singly or in groups, in real-time event streams or in stored datasets. In this section we will first apply standard SQL to example network management problems. ANQL will be introduced to reduce certain complexities of using SQL typical protocol processing situations, and to make available further application-specific language features.

## 3.1   SQL Examples

Let us apply the RDBMS model to a simple problem in network management. Suppose we wish to extract a trace of the source and destination IP addresses of every packet in a sequence of packets. Assume that the packet data collected at some prior time, and is stored in RDBMS tables as described in Sect. 2. In this example we need only a single table, **ipv4**, to hold IPv4 header data from each packet [1]. Figure 1 shows an SQL query on this data (written as a database *view*). The IPv4 protocol header address fields are stored in database fields named **saddr** and **daddr**, which are accessed using the SQL notation **ipv4.saddr** and **ipv4.daddr** to clarify the data source involved.

```
CREATE VIEW ip_packets
AS
SELECT ipv4.saddr, ipv4.daddr
FROM ipv4
```

**Fig. 1.** SQL statement for extracting IP addresses.

Let's consider a more complex (yet not atypical) operation, such as decoding RIP [7] packets. For this example, we will use a RIP implementation operating in an ASP EE [2] virtual network topology running on the ABone [1] [2]. We want to extract the source and destination virtual addresses and the RIP command

---

[1] In this paper, protocol data will be represented by a database table with the same name as the protocol, except for case

[2] The ASP EE supports the protocols identified in the example as VN, VT, and ASP. The ABone supports the protocol identified as ANEP, as well as the constants used to match UDP port 3322 and ANEP TypeID 135.

from each packet. Furthermore, we have to filter these packets out of a general packet stream that may contain many types of packets, only some of which are of interest to us. Figure 2 shows one possible SQL command to do this.

```
CREATE VIEW rip_packets1
AS
SELECT vn.saddr, vn.daddr,
       decode(rip.command,
              1, "rip_request",
              2, "rip_response",
                 "rip_other") command_name
FROM ipv4, udp, anep, vn, vt, asp, rip
WHERE ipv4.protocol = 17
  AND  udp.packet_id = ipv4.packet_id
  AND  udp.dport     = 3322
  AND anep.packet_id = ipv4.packet_id
  AND anep.typeid    = 135
  AND   vn.packet_id = ipv4.packet_id
  AND   vt.packet_id = ipv4.packet_id
  AND   vt.dport     = 520
  AND  asp.packet_id = ipv4.packet_id
  AND  asp.aaname    = "rip"
  AND  rip.packet_id = ipv4.packet_id
```

**Fig. 2.** SQL statement for examining RIP packets.

The **SELECT** clause extracts the virtual source and destination addresses (**vn.saddr** and **vn.daddr**) and a text representation of the RIP command (using **decode(...)**, which is a function for mapping data values that is found in some dialects of SQL). SQL has the expressive power to handle this example, but the large number of conjunctions in the **WHERE** clause is clumsy.

# 4   ANQL Examples

Figure 3 shows the same example, using ANQL. Compared to SQL, the primary changes are that the **FROM** clause contains a protocol specification (see Appendix A) for particulars) and the **WHERE** clause is devoid of the implicit comparisons on **packet_id**. **CREATE ACTIVE FILTER** states that this particular ANQL statement is creating an Active Network packet filter (as opposed to a database view). **USING NETIOD** specifies that the packets are to be acquired in real time through Netiod[1], a program that provides system-independent access to packet flows on Unix systems.

The ANQL **WHERE** clause contains the information that a NodeOS [6] channel specification carries in the address specification and optional demux

```
CREATE ACTIVE FILTER rip_packets2
AS
SELECT vn.saddr, vn.daddr,
       decode(rip.command,
                1, "rip_request",
                2, "rip_response",
                   "rip_other") as status
FROM "if/ipv4/udp/anep/vn/vt/asp/rip"
WHERE ipv4.protocol = 17
  AND   udp.dport   = 3322
  AND anep.typeid   = 135
  AND    vt.dport   = 520
  AND   asp.aaname  = "rip"
USING NETIOD
```

**Fig. 3.** ANQL statement for examining RIP packets.

specifications. Unlike the NodeOSpositional notation, the **protocol.field** notation in ANQL is easily expandable and relatively self-documenting.

There are still some redundancies in ANQL. In Figure 3, WHERE ipv4.protocol = 17 says that the protocol after IPv4 should be UDP. Although ANQL could have automatically added this constraint to the **WHERE** clause based on the contents of the **FROM** clause, ANQL takes the approach that all magic numbers that define protocol relationships should be shown explicitly; this is a user interface issue more than an essential property of the ANQL language itself.

The SQL database language, as implemented by commercial database vendors, can perform data transformations in the **SELECT** and **WHERE** clauses. In addition to the usual numerical and logical operations, SQL implementations often support string manipulation, date/time calculation, table-based data value remapping, and other non-numeric operations

The initial ANQL implementation supports many of the data transformation capabilities that are typical of SQL, and adds specialized functions that are appropriate to the area of Active Networking. Figure 4 extends the **rip_packets2** filter by calling the **vnetToHost(...)** function to convert a virtual network address into a host name. The **AS** syntax in the **SELECT** clause is used to rename the computed output fields. The result is a tuple with fields named (**node, node2, property, status**), which meets the input requirements of the network packet visualizer [9] for which this script was used to filter packets in real time.

SQL provides a **GROUP BY** clause to direct the computation of summary statistics. In ANQL, it can be used to to summarize over time or over the attributes of the packets. For example, Fig. 5 extracts the bandwidth consumed by each RIP command type, in octets per second averaged over 10 minute intervals. The **interval(...)** function shown here is an example of an ANQL extension to

```
CREATE ACTIVE FILTER rip_packets3
AS
SELECT vnetToHost(vn.saddr) AS node,
       vnetToHost(vn.daddr) AS node2,
              "rip-packet!" AS property,
       decode(rip.command,
              "1", "rip_request",
              "2", "rip_response",
                   "rip_other") as status
FROM "if/ipv4/udp/anep/vn/vt/asp/rip"
WHERE ipv4.protocol = 17
   AND   udp.dport = 3322
   AND anep.typeid = 135
   AND   vt.dport  = 520
   AND   asp.aaname = "rip"
USING NETIOD
```

**Fig. 4.** ANQL for extracting and reformatting RIP data.

SQL that was created to better adapt the language to the packet processeing domain, in this case by simplifying the manipulation of time intervals.

```
CREATE ACTIVE FILTER rip_bandwidth
AS
SELECT rip.command, sum(ipv4.length)/(10*60) AS bandwidth
FROM "if/ipv4/udp/anep/vn/vt/asp/rip"
WHERE ipv4.protocol = 17
   AND   udp.dport = 3322
   AND anep.typeid = 135
   AND   vt.dport  = 520
   AND   asp.aaname = "rip"
GROUP BY rip.command, interval(sysdate, 10, "MI")
USING NETIOD
```

**Fig. 5.** ANQL for calculating RIP bandwidth.

# 5   Applications of ANQL

ANQL is an application-specific language. As such, it can express application-specific requests more compactly than a more general purpose language (such as general Java byte code). Thus, for many purposes ANQL commands can be included in packets without sacrificing a lot of bandwidth.

The primary applications envisioned for ANQL are in the control or manangement planes in a network. In these applications, ANQL scripts function as commands to control or management application programs (which might be implemented as active applications or as inactive ones), and are not strictly speaking active packets. On the other hand, ANQL scripts may be used to create distributed sessions for management reporting (such as by creating a tree of ANQL scripts passing summary information towards a common root), and in this application ANQL scripts may operate as active packets.

## 5.1  Filtering Packets

The initial uses for ANQL have been to filter packets from remote packet intercept points. A single ANQL processor runs on a central system, operating on remotely gathered data or even on stored packet traces; the examples in Sect. refsec:anqlExamples illustrate this capability. An ANQL-based filter may also be distributed to packet collection points in an Active Network topology, with little or no change in the filter itself. Thus, the ANQL-based filters can scale beyond the processing limits of a centralized filter.

## 5.2  Summary Statistics

ANQL scripts can also be used to collect summary statistics on packet flows in an Active Network. In a small network, summary statistics can be computed by a single, central node that collects intercepted packet streams from all other nodes in the the topology. In a larger network, the ANQL script can be distributed to a selection of nodes in the topology, with little change in the ANQL script. Beyond that, ANQL scripts can be distributed to compute summary statistics in a heirarchical computation tree, with little change to the ANQL script. This provides a high degree of scalability.

ANQL, as a dialect of SQL, provides **MIN**, **MAX**, **SUM**, **AVG**, **STDDEV**, and other descriptive statistics in the language. Since many summary statistics of packet traffic are supported by ANQL itself, that code does not need to be written into Active Applications (AAs) that require the summary data. This greatly simplifies the implementation and maintenance of these AAs.

## 5.3  Triggering Actions

ANQL scripts can be used to trigger actions in an Active Network. Figure 6 has an active trigger that uses the SQL "group by" and "having" clauses (in their ANQL incarnations) to issue restart commands for neighboring nodes that have sent more than 3 erroneous RIP packets in a 10-minute interval. The restart action is created as a text string containing the word "restart" and the virtual network address of the failed node; presumably this is a command to the application using ANQL.

It would not be difficult to use ANQL to generate operating system commands (shell commands) for execution on local or remote nodes. Report writing scripts for early versions of Oracle's SQL tools often operated in an analogous fashion. The security implications of this technique are outside the scope of this paper.

```
CREATE ACTIVE TRIGGER restart_neighbor
AS
SELECT "restart "||vn.saddr AS action
FROM "if/ipv4/udp/anep/vn/vt/asp/rip"
WHERE ipv4.protocol = 17
  AND  udp.dport = 3322
  AND  anep.typeid = 135
  AND   vt.dport  = 520
  AND  asp.aaname = "rip"
  AND  rip.command != "1"
  AND  rip.command != "2"
GROUP BY vn.saddr, interval(sysdate, 10, "MI")
HAVING count(*) > 3
USING NETIOD
```

**Fig. 6.** ANQL for restarting failing neighbors.

# 6   Relation to the NodeOS Channel Spec

The ANQL **FROM** clause contains a NodeOS-like protocol specification. The ANQL **WHERE** clause contains the information that is carried in the NodeOS address specification and optional demux specification(s). Unlike the positional notation used in the NodeOS address specification, ANQL's **protocol.field** notation is easily expandable and relatively self-documenting. As a result, it is expected that ANQL packet filters will be easier to program and maintain than equivalents programs written directly in NodeOS channel specifications.

# 7   Implementation and Efficiency

ANQL is currently implemented using a general expression interpreter [8] written in Java. This implementation decision provided a high degree of functionality at the cost of runtime efficiency. The design of ANQL does not preclude compilation to Java or C, or even to machine code for use on specialized network protocol processors. The ANQL language (excluding the possible procedural extensions discussed elsewhere in this paper) is non-procedural; this quality should make it easier to compile ANQL to platform-specific code.

There are certain optimizations that could be applied to the present implementation of ANQL. For example, when processing protocol headers, all possible protocol fields (**ipv4.saddr**, **ipv4.daddr**, etc.) are extracted from the packet and saved in a Java util.Hashtable. It should be possible to analyze the protocol fields used in the ANQL expressions and extract only the fields that are specifically needed.

## 8  Capsule Applications of SQL-like Languages

*Capsule languages* are languages used to write Active Network programs that are carried directly in active packets; one example is [10]. Due to packet size limitations in typical computer networks, capsule languages face a difficult tradeoff between brevity and functionality. The focus of a capsule language, operating in the *data plane* of an active network, tends to center on the selection of node resources to match the processing requirements of the packet. ANQL, as described here, focuses on the extraction of data from individual packets or groups of packets; it is not really suitable for use as a capsule language.

It is possible to envision a role for an alternative SQL-like language as a pure capsule language. Instead of filtering and manipulating the contents of one or more packets, the language would support tasks such as selecting among protocol processing modules in a node or selecting from a set of nodes for packet forwarding, using a combination of the available set of resources at each node traversed by the capsule and the data carried in the packet itself. Figure 7 has a simple example that illustrates this approach. The sample program selects the node with the shortest queue as the next node for the current packet (ties are broken by selecting the node with the lowest address):

```
SELECT n.addr next_node
FROM neighbor_nodes n
WHERE n.queue_length in (
    SELECT min(n2.queue_length)
    FROM neighbor_nodes n2
)
ORDER BY n.addr
```

**Fig. 7.** SQL for active packet routing.

As described in Sect. 9, the SQL-like capsule language could be augmented with a procedural extension. The result, a computationally complete language, could be compiled into byte codes and used as an Active Network capsule language: the source code would be high-level and self-documenting, while the compiled byte codes could easily be as compact as Java's, if not more so.

## 9  Further Research

The ANQL language continues to grow to meet the requirements of network control and management, and of Active Networking in particular. Additional functions are being written for use in ANQL expressions, and range of protocols that can be parsed by ANQL continues to expand.

ANQL is similar to Oracle's implementation of SQL, which is well-known and relatively accessible. There are other languages related to ANSI SQL with interesting features, such as time comparisons extensions, that might be desirable in the Active Networks domain. It could be useful to incorporat these features into ANQL.

Oracle Corp. has created a procedural language extension to SQL, PL/SQL. There is also an ANSI SQL procedural extension for SQL. A procedural extension is feasible for ANQL; such an extension could be used as a portable, computationally complete active packet language.

We would like to investigate a SQL-based capsule language, as described in Sect. 8. This approach represents an interteting tradeoff between semantic expresiveness, representational compactness, and implementation complexity, compared to prior efforts[10].

A nonprocedural language, such as ANQL, may also have certain advantages when analyzing the safety of statements in the language. It would be interesting to pursue the safety and security properties of ANQL.

ANQL could replace the NodeOS channel specification in some programs. The translation from ANQL to lower-level channel mechanisms is a one-time operation, which may not be significant when amortized over the lifetime of a packet flow. If necessary, the translation could be cached for reuse at runtime (if the same channel specification is opened multiple times in the lifetime of an EE) or precompiled (in the style of commercial SQL precompilers) into the source code of an active application.

## 10   Related Work

ANQL uses Netiod [1], an operating-system neutral packet filter interface, to implement portions of its packet acquisition and filtering process. ANQL could be implemented on top of the BSD Packet Filter [11] or on top of a more system-specific packet filtering mechanism, such as IPCHAINS or IPTABLES on Linux. In all of these cases, ANQL would extend the functions of the lower-level packet filter with ANQL's SQL-based syntax and semantics.

NNStat [3] [4] provides a low-level packet filter, a remote packet collection facility, and several higher-level data analysis capabilities. ANQL and NNStat use similar **protocol.field** naming conventions, but NNStat is much more concerned with the lower-level details of the packet filter implementation than is ANQL, which focuses on higher-level issues.

## 11   Summary and Conclusions

The Active Networks Query Language applies the expressive power of SQL to the domain of packet network control and management. This increase in expressive power makes ANQL easier to use than NodeOS channel specifications in many applications; furthermore, many common operations, such as computing summary statistics about packet traffic, can be expressed directly in ANQL rather than requiring custom code in each application that needs them. The initial implementation of ANQL is interpretive, but ANQL could alternatively be compiled directly into Java, C, or machine code for efficient execution.

## References

[1] Steven Berson, Steven Dawson, and Robert Braden. Evolution of an Active Network Testbed. In *DARPA Active Networks Conference & Exposition*, pages 446–465, May 2002.

[2] Bob Braden, Alberto Cerpa, Ted Faber, Bob Lindell, Graham Phillips, and Jeff Kann. ASP EE: An Active Execution Environment for Network Control Protocols. http://www.isi.edu/active-signal/ARP, 1999.

[3] Robert T. Braden. A Packet Monitoring Program. Technical report, USC/Information Sciences Institute, March 1990.

[4] Robert T. Braden and Annette L. DeSchon. NNStat: Internet Statistics Collection Package: Introduction and User Guide. Technical report, USC/Information Sciences Institute, August 1988.

[5] E. F. Codd. A Relational Model of Data for Large Shared Data Banks. *Communications of the ACM*, 13(6):377–387, June 1970.

[6] AN Node OS Working Group. NodeOS Interface Specification. http://www.cs.princeton.edu/nsg/papers/nodeos.ps, January 2000.

[7] G. Malkin. RIP Version 2. RFC 2453, November 1998.

[8] Craig Milo Rogers. The ABoneShell. http://www.isi.edu/abone/ABoneShell.html.

[9] Craig Milo Rogers. ABoneMonitor Packet Visualizer Demo. DANCE 2002, San Francisco, CA, May 2002.

[10] B. Schwartz, W. Zhou, A. W. Jackson, and et. al. Smart Packets for Active Networks. Technical report, BBN Technologies, January 1998.

[11] Van Jacobson Steven McCanne. The BSD Packet Fitler: A New Architecture for User-level Packet Capture. In *Proceedings of the Winter 1993 USENIX Conference*, pages 259–270, January 1993.

[12] D. L. Tennenhouse and D. J. Wetherall. Towards an Active Network Architecture. In *http://www.tns.lcs.mit.edu/publications/ccr96.html*, 1996.

[13] X3.135. Database Language SQL. Technical report, ANSI, 1992.

[14] J. Zander and R. Forchheimer. SOFTNET - An Approach to High Level Packet Communications. In *Proceedings of the AMRAD Conference*, 1983.

## A   The ANQL Language

The initial implementation of ANQL closely resembles SQL.

```
CREATE ACTIVE [FILTER | TRIGGER] <name>
AS
SELECT <selectExpr> [AS <fieldName>] [, ...]
FROM <protocolSpec>
[WHERE <whereExpr>]
[GROUP BY <groupExpr>]
[HAVING <havingExpr>]
[ORDER BY <orderExpr>]
[USING NETIOD]
```

The expressions may contain logical operators, arithmetic operators, comparisons, functions, etc. In general, these behave as they would in SQL. ANQL's functionality can be easily increased because the ANQL interpreter can be dynamically extended at runtime by loading Java code to implement new operators and functions.

The **WHERE** clause selects records before grouping and the **HAVING** clause selects record groups after grouping, as in SQL. ANQL's **FROM** clause uses a protocol specification that is similar to the NodeOS channel specification, such as:

```
FROM "if/ipv4/udp/anep/vn/vt/asp/rip"
```

The protocol names are separated by slashes. Each protocol, when matched against an incoming packet, makes certain data fields available for use in expressions in the ANQL statement. To repeat a protocol, such as in IP/IP tunneling, a colon-separated suffix can be attached to each protocol name to disabiguate the protocol layers in ANQL expressions; this suffix mechanism is similar to the table name alias feature of some SQL implementations. Example:

```
SELECT ipv4:2.saddr
FROM "if/ipv4:1/ipv4:2/udp/anep/vn/vt/asp/rip"
```

Here is another example, showing IPv4 tunneling within a UDP envelope. In this case, only the outer IPv4 and UDP headers have been given a special suffix, and the **SELECT** extracts the inner IPv4 header:

```
SELECT ipv4.saddr
FROM "if/ipv4:outer/udp:outer/ipv4/udp/anep/vn/vt/asp/rip"
```

In addition to the protocols and fields mentioned in the **FROM** clause of an ANQL statement, certain special fields may be available in an ANQL expression. For example, when using Netiod to intercept packets, **netiod.raddr** may be the IP address of the Netiod instance that intercepted the packet, and **netiod.timestamp** may be the Netiod-supplied timestamp for when the packet was intercepted.

As of this writing, ANQL parsers have been implemented for the following network protocols:

```
General: IPv4, RDP, TCP, UDP
Active Networking: ANEP, Netiod
ASP EE: AASpec, UI, VN, VT, VTS
```

As an example, the IPv4 fields that are available for use in ANQL expressions are:

```
ipv4.length, ipv4.df, ipv4.mf, ipv4.ttl, ipv4.tos, ipv4.protocol,
ipv4.saddr, ipv4.daddr
```

# Predictable, Lightweight Management Agents*

Jonathan T. Moore[1], Jessica Kornblum Moore[1], and Scott Nettles[2]

[1] Computer and Information Science Department
University of Pennsylvania
{jonm,jkornblu}@dsl.cis.upenn.edu
[2] Electrical and Computer Engineering Department
The University of Texas at Austin
nettles@ece.utexas.edu

**Abstract.** In this paper we present an *active*, or programmable, packet system, SNAP (Safe and Nimble Active Packets) that can be used to provide flexible, lightweight network management with predictable resource usage. SNAP is efficient enough to support both centralized polling and mobile agent approaches; furthermore, its resource predictability prevents malicious or erroneous agents from running amok in the network.

## 1 Introduction

Mobile agents have often been proposed as a solution for network management [5]. By moving computation away from a central network operations center (NOC) via mobile code, these schemes use distribution to achieve scalability [4, 5, 6, 17, 20]. Furthermore, having a programmable management platform in theory makes it easier to customize management applications to particular networks or to quickly take new management actions (for example, a response to a new virus or denial-of-service attack).

If mobile agents are such a panacea for distributed management, why do they not appear more frequently in actual management systems? We believe there are two reasons: existing systems are too heavyweight, and network managers fear losing control over runaway agents.

Several researchers have explicitly considered how the performance of agents compares to SNMP and similar approaches. Rubinstein and Duarte [20], for example, simulate the performance of an agent designed to gather values from several nodes and compare it to SNMP on a simple topology. Their simulations show that for several metrics when a large number of nodes are visited (over 200), their agent outperforms SNMP. However, for a small number of nodes, SNMP is superior; because their agent requires about 5,000 bytes to be transmitted, the mobile agent approach can overcome its constants of proportionality only for the largest of networks. Baldi and Picco [3] develop a set of analytical models concerning the performance of several approaches, including both centralized polling and mobile agents and are thus able to characterize when one

---

* This work was supported by DARPA under Contract #N66001-96-C-852 and by the NSF under Contracts ANI 00-82386 and ANI 98-13875.

J. Sterbenz et al. (Eds.): IWAN 2002, LNCS 2546, pp. 111–119, 2002.

technique will perform better than another. They then examine a specific case study involving SNMP and Java Aglets [13]. Here the Java code is again about 5,000 bytes in size, with a similar impact on the trade-offs compared to SNMP.

A further problem concerns bounding resource usage, which is always an issue where mobile agents are concerned. Indeed, many agent platforms suffer from the problem of runaway agents that are difficult to track down and stop once unleashed [27]. Network operators are extremely hesitant about surrendering control to agents whose behavior is not entirely predictable.

In this paper, we describe the use of an *active* (or, programmable) packet system, SNAP (Safe and Nimble Active Packets) [15], for use in network management. SNAP packets contain code that is executed at SNAP-aware routers, thus adding significant flexibility over IPv4 or simple client-server protocols. The SNAP language has been designed to provide lightweight execution as well as predictable resource usage. As a result, SNAP is an ideal candidate for realizing the mobile agent approach to network management.

We begin in Section 2 with a brief overview of SNAP, stressing its security features and providing a micro-benchmark demonstrating its lightweight implementation. Then in Section 3, we illustrate the detection of a Distributed Denial-of-Service (DDoS) attack using special SNAP monitoring agents. We discuss related work in Section 4 before concluding and suggesting future work in Section 5.

## 2    SNAP: Predictable, Lightweight Mobile Agents

SNAP is an active packet language designed to address the open problem of providing a flexible programming language with high performance, yet safe (and in particular, safe with respect to system resources), execution. In this subsection, we present merely a high level overview; the reader is referred to [15] for more details.

SNAP is a simple stack-based bytecode language designed to require no unmarshalling (and in most cases very little or no marshalling) and to permit in-place execution. The packet contains code, heap, and stack segments; the stack is the last segment in the packet, allowing us to execute SNAP *in-place* in a network buffer, as the stack can grow and shrink into the available space at the end of the buffer (the heap can also grow down from the end of the buffer if needed). By design, SNAP does not require a garbage collector.

SNAP is indeed a low-level, assembly-style language (see Figures 1 and 2 for examples), making tightly hand-tuned packet programs possible. However, for programming convenience, a compiler [10] exists to translate the higher-level, strongly-typed active packet language PLAN into SNAP.

In its current implementation, SNAP is carried in an IPv4 packet with the Router Alert [11] option. Thus, legacy routers will simply forward a SNAP packet toward its destination. On the other hand, SNAP-aware routers can detect the router alert, check the IP protocol field to see that the packet contains SNAP, and then call the SNAP interpreter. Kernel and user-space implementations exist for Linux systems, and an interpreter has also been built for IBM's PowerNP 4GS3 network processor [12].

A key aspect of SNAP's design is its model of predictable resource usage. In particular, a single packet execution on a node is guaranteed to use time and space proportional to the packet's length. This guarantee is derived from two facts: all SNAP instructions execute in (possibly amortized) constant time and space, and all branches must move forward. Thus, unlike other approaches based upon watchdog timers or memory limits, SNAP routers know that incoming SNAP programs will behave reasonably with respect to resources *without even having to examine them*. Furthermore, this means that SNAP programmers can know that their programs will not be unexpectedly terminated; if they can express their programs in SNAP, those programs are resource safe.

In addition, each packet has an associated *resource bound* (RB) field that is decremented for every network hop, including sending to the current node via a loopback interface. This loopback behavior can be used to emulate backward branches at the cost of 1 RB. Furthermore, a packet must donate some of its RB to any child packets it spawns (conservation of resource bound). As a result, we can examine the initial length of a SNAP program and its initial RB count, and place a strict upper bound on the amount of network resources it or its descendants can ultimately use.

Because SNAP cannot express infinite loops, nor can SNAP packets roam around the network indefinitely, it is not possible to write a stationary daemon-style agent that "settles in" at a given node to monitor it, nor can we write permanently circulating "sentry" agents. Stationary agents are more appropriately designed using existing approaches like AgentX [7] and RMON [25]. Long-running agents that traverse a given circuit repeatedly can be composed from multiple SNAP agents, each of which traverses the circuit a fixed number of times before reporting back into the NOC, which then sends out a new agent. Our example DDoS detection application in Section 3 would be naturally constructed this way. The benefit, of course, is that to control incorrect agents, we can simply stop injecting new agents at the NOC; all outstanding agents will then fairly quickly run out of resources and die.

For added security, access to management functions may be protected by strong authentication, in the style of PLAN [9]. Here, there is some minimal set of "safe" services available to all packets without authentication. A packet can carry cryptographic credentials that can be presented to the node in exchange for an expanded service namespace—access to additional services. This style of *namespace security* allows packets to only pay the costs of cryptographic authentication on an as-needed basis. Although we have not yet implemented this service for SNAP, it would be straightforward: the credentials could be carried in the packet's heap as a byte array and then handed to an "auth" service that would side-effect the service namespace for the packet.

```
          forw                            ; move towards destination
          bne done                        ; if returning to source, branch
          push "interfaces.ifNumber.0"    ; MIB variable
          calls "snmpget"                 ; invoke service
          push 1                          ; push 1 to indicate return
          getsrc                          ; retrieve source address
          forwto                          ; forward back toward source
done:
          demuxi <portnum>                ; deliver payload
```

**Fig. 1.** SNAP program for the polling micro-benchmark.

## 2.1   Micro-benchmark

In this section, we support our claim that SNAP can provide a lightweight network monitoring system. Our experiments were performed on a two PCs, called *hera* and *athena*; each machine has a Pentium III (Coppermine) 1 Ghz CPU, 256 MB of RAM, and a SuperMicro Super 370 DE6 motherboard with on-board Intel Speedo3 100 Mbps Ethernet card. The cluster runs RedHat Linux 7.3, with kernel version 2.4.18-5. Both machines are on the same LAN, switched by an Asanté Fast100 Ethernet hub. The measurements we present are the median of 21 trials.

In our first test, we ran the snmpd from ucd-snmp version 4.2.5 on *hera*, and ran a client program on *athena* to retrieve the interfaces.ifNumber.0 MIB variable from *hera*. The median latency was 575 $\mu$s (the ping latency between the machines accounts for 148 $\mu$s of this).

In our second test, we ran our user-space snapd on both *hera* and *athena*, and injected the SNAP program[1] shown in Figure 1 from *athena*. The program proceeds to *hera*, queries the MIB variable via the "snmpget" service, and then returns back to *athena*. The median latency for this program was 660 $\mu$s, an extra overhead of only 85 $\mu$s (15%) over plain snmpd above. However, this overhead is exaggerated by the fact that the implementation of the "snmpget" service simply contacts the local snmpd via a network socket; this overhead could be avoided in an integrated SNAP+SNMP daemon, such as we discuss in future work (Section 5).

## 3   Application Example: DDoS Detection

We now present a concrete example using SNAP as a lightweight mobile agent platform for network management. Distributed denial-of-service (DDoS) attacks are an increasing problem, targeting well-known e-commerce and government sites; easy to use tools for carrying out these attacks [1], are becoming widely available. For e-commerce sites, response time for such attacks is critical, as

---

[1] For an explanation of SNAP semantics, the reader is referred to [15].

serious revenue losses can accrue during down-time, not to mention the impact on customer satisfaction. The first step of a response is detecting the attack in the first place. In this section, we describe a DDoS detection mechanism using SNAP.

To detect a DDoS attack, we can measure the amount of incoming traffic $T$ into our administrative domain. Generally, as a network manager, we will have some traffic threshold $T_{alarm}$; if incoming traffic exceeds $T_{alarm}$, we want to sound an alarm and take action. We need to query incoming octet counts on multiple interfaces of multiple nodes. The usual centralized polling approach will quickly overwhelm the NOC with management data as the number of managed nodes grows, so this solution does not scale. The key to scalability lies in being able to distribute the computation of $T$. Fortunately, this particular computation can be performed incrementally, making it especially well suited for the use of lightweight active packets.

## 3.1  SNAP Surveyors

Figure 2 shows the SNAP "surveyor" programs that we use to address the scaling problem. This packet program carries a list of nodes to query and visits each in sequence. At each node $i$ it queries the MIB variable `interfaces.ifTable.ifEntry.ifInOctets` for the external interface and keeps a running sum, $T_{sum}$. Once all nodes have been visited, the surveyor returns to the NOC and reports the current value of $T$. The algorithmic intuition is the following: with $n$ nodes to manage, a centralized polling approach requires $O(2n)$ network hops (out to each node and back), whereas the surveyor approach requires $O(n)$ hops. Perhaps more importantly, in the centralized approach, all $O(2n)$ hops involve the NOC, whereas in the surveyor approach, only 2 hops involve the NOC. Thus, not only is the network traffic reduced, but it is also distributed.

The surveyor program consists of just 21 instructions, for a total of 84 bytes of code in a SNAP packet. This leaves significant room in the packet for carrying accumulated data and/or addresses of nodes to visit. Even with a maximum transmission unit (MTU) as small as 256 bytes, there are still 128 bytes of room left over in the packet after headers and code (enough to visit 32 nodes, assuming 4 bytes of accumulated data as in the above example). With more realistic autonomous domain MTUs of 1500 bytes, one packet could easily visit over 300 nodes.

## 3.2  Discussion

The specific example presented here is just one of a class of *distributed threshold detection* problems; Raz and Dilman [18] point out several such problems, including monitoring general network traffic, Web mirror loads, software licenses, bandwidth brokerage, and denial-of-service attacks. In each case, we want to know whether some global network threshold has been exceeded.

Raz and Dilman's approach, efficient reactive monitoring, apportions some "ration" of the global threshold to each monitored node; the node monitors its

```
main:
        forw                ; get to next hop
        bne     athome      ; if homeward flag set, just deliver data

        ;; else, we need to update load sum
        push  "interfaces.ifTable.ifEntry.ifInOctets.5"
        calls "snmpget" ; get current octet count
        add                 ; running sum

        ;; any more nodes to visit?
        pull    1           ; get n (number of remaining nodes)
        bez     gohome      ; if out of addrs, go home

        ;; re-arrange stack state in preparation for transit
        pull    2           ; get next node's address
        pull    2           ; get n
        subi    1           ; n--
        store   4           ; put new n over old next hop
        pull    1           ; pull load sum
        store   3           ; put load sum over old n
        push    0           ; still more hops to go; unset homeward flag
        store   2           ; put flag over old load sum
        forwto              ; move on to next hop

gohome:
        push    1           ; set homeward-bound flag
        getsrc              ; find out where home is
        forwto              ; go there

athome:
        getspt              ; get port number for delivery
        demux               ; deliver octet total
```

**Fig. 2.** SNAP "surveyor" program

own state and, if the ration is used up, triggers an alarm to the NOC, which then issues a global poll. If none of the nodes exceed their ration, then the global usage cannot have exceeded the global threshold.

This approach can be adapted to use SNAP surveyors without requiring extra code to be installed at the monitored nodes (except for the one-time installation of a SNAP interpreter). Furthermore, the surveyor can make use of domain-specific knowledge to shortcut its route: the surveyor may be able to determine, based on its current incremental result, the number of remaining nodes to visit, and an upper bound on the queried value, that the overall threshold will not be exceeded[2].

---

[2] A centralized poll could also short-circuit its search, but as we noted earlier, this requires on the order of twice as many network hops as the surveyor approach.

# 4  Related Work

There is a significant body of work concerning mobile agents in network management [2, 4, 6, 16, 24, 17, 19, 21, 22, 26]. Bieszczad *et al.* provide a survey [5] that identifies a number of areas in which mobile agents are applicable to network management, including network modeling, fault management, configuration management, and performance management.

Existing mobile agent systems [4, 16, 24, 17, 19] geared toward network management tend to be based largely upon Java. Unfortunately, there is no good way to bound the resources consumed by a Java agent, as infinite loops can be expressed, yet Hawblitzel *et al.* have shown that it is unsafe to simply terminate runaway threads in the JVM [8].

The IETF ScriptMIB [14] provides an SNMP-based interface for installing and running scripts, although it does not specify a particular script programming language. Furthermore, multiple SNMP round-trips are necessary to set up and invoke a new script, whereas SNAP-based agents are more "light on their feet," being self-contained.

Perhaps the most closely related project to ours is the Smart Packets project from BBN [23]. Indeed, their system closely resembles ours in having a byte-code interpreter for active packets. The main difference from our work is their approach to resource control. Smart Packets rely on instruction counters and memory limits to prevent packets from consuming too many local resources, whereas SNAP can provide the same guarantees via language design. In the end, this impacts programmer convenience: with Smart Packets, it is possible that a packet program may accidentally and unpredictably exceed its resource allotment and be prematurely terminated, whereas SNAP programs will always run to completion (barring other sorts of errors). One other important difference is that Smart Packets cannot direct themselves; they are sent from a source to a destination, and may execute on intermediate nodes, but may not deviate from the original path. As a result, Smart Packets do not offer quite the agility needed for truly mobile agents. Finally, no performance data is available to determine whether the Smart Packets execution environment is lightweight or not.

# 5  Conclusions and Future Work

We have described an active packet system, SNAP, and have argued that it can be used to add flexible mobile agent capabilities wherever SNMP is already used. We have shown experimentally that SNAP execution overheads are small compared to SNMP: thus SNAP offers the flexibility of mobile agents with the efficiency of standard centralized polling. Furthermore, our language design guarantees that SNAP agents have predictable (and finite) resource usage. Finally, we have presented an example application, using SNAP to detect distributed denial-of-service attacks (DDoS).

The main thrust for future work revolves around providing a generic SNMP service interface for SNAP. Walter Eaves of University College London has already added an SNMP service interface to a user-space SNAP implementation.

In this system, services send appropriate SNMP requests to a separate SNMP daemon on the same host. Willem de Bruijn at the Leiden Institute of Advanced Computer Science is currently merging a SNAP user-space implementation with the `net-snmp` SNMP daemon, creating a single program with a MIB backend and two frontends, one for vanilla SNMP and one for SNAP. The resulting program will be a SNAP-enabled drop-in replacement for a standard `snmpd`.

## Acknowledgments

The authors would like to thank Michael Hicks and Aaron Marks for comments on early drafts of this paper, Sotiris Ioannidis for useful discussions about using active network technology for network monitoring, and Walter Eaves for discussions about an SNMP service interface for SNAP and for implementing such an interface for the user-space version of SNAP. We would also like to thank the anonymous reviewers and Rolf Stadler for their useful comments.

## References

[1] Denial of service tools. CERT Advisory CA-1999-17, December 1999.

[2] M. Baldi, S. Gai, and G. Picco. Exploiting Code Mobility in Decentralized and Flexible Network Management. In *Proceedings of the First International Workshop on Mobile Agents*, April 1997.

[3] Mario Baldi and Gian Pietro Picco. Evaluating the tradeoffs of mobile code design paradigms in network management applications. In *Proceedings of the 20th International Conference on Software Engineering (ICSE'98)*, April 1998.

[4] C. Baumer and T. Magedanz. The Grasshopper Mobile Agent Platform Enabling Shortterm Active Broadband Intelligent Network Implementation. In *Proceedings of the First International Working Conference on Active Networks (IWAN'99)*, June/July 1999.

[5] A. Bieszczad, T. White, and B. Pagurek. Mobile Agents for Network Management. *IEEE Communications Surveys*, 1(1), September 1998.

[6] M. Breugst and T. Magedanz. Mobile Agents—Enabling Technology for Active Intelligent Network Implementation. *IEEE Network*, 12(3):53–60, May/June 1998.

[7] M. Daniele and B. Wijnen. Agent Extensibility (AgentX) Protocol, Version 1. RFC 2741, IETF, January 2000.

[8] C. Hawblitzel, C.-C. Chang, G. Czajkowski, D. Hu, and T. von Eicken. Implementing Multiple Protection Domains in Java. In *Proceedings of the 1998 USENIX Annual Technical Conference*, June 1998.

[9] M. Hicks and A. Keromytis. A Secure PLAN. In *Proceedings of the First International Working Conference on Active Networks (IWAN'99)*, June/July 1999.

[10] M. Hicks, J. Moore, and S. Nettles. Compiling PLAN to SNAP. In *Proceedings of the IFIP-TC6 Third International Working Conference on Active Networks (IWAN'01)*, pages 134–151, September/October 2001.

[11] D. Katz. IP Router Alert Option. RFC 2113, IETF, February 1997.

[12] A. Kind, R. Pletka, and B. Stiller. The potential of just-in-time compilation in active networks based on network processors. In *Proceedings of the 5th Workshop on Open Architectures and Network Programming (OPENARCH'02)*, pages 79–90, June 2002.

[13] Danny B. Lange. Java aglet application programming interface (J-AAPI) white paper—draft 2, 1997.

[14] D. Levi and J. Schoenwaelder. Definitions of Managed Objects for the Delegation of Management Scripts. RFC 3165, IETF, August 2001.

[15] J. Moore. *Practical Active Packets*. PhD thesis, University of Pennsylvania, 2002.

[16] Objectspace, Inc. Voyager application server 4.0 datasheet, October 2000.

[17] A. Puliafito and O. Tomarchio. Using Mobile Agents to Implement Flexible Network Management Strategies. *Computer Communications Journal*, 23(8), April 2000.

[18] D. Raz and M. Dilman. Efficient Reactive Monitoring. In *Proceedings of the 20th Annual Joint Conference of the IEEE Computer and Communications Societies*, April 2001.

[19] D. Raz and Y. Shavitt. An Active Network Approach for Efficient Network Management. In *Proceedings of the First International Working Conference on Active Networks (IWAN'99)*, June/July 1999.

[20] M. Rubinstein and O. Duarte. Evaluating Tradeoffs of Mobile Agents in Network Management. *Networking and Information Systems Journal*, 2(2), 1999.

[21] A. Sahai, C. Morin, and S. Billiart. Intelligent Agents for a Mobile Network Manager. In *Proceedings of the IFIP/IEEE International Conference on Intelligent Networks and Intelligence in Networks (2IN'97)*, September 1997.

[22] C. Schramm, A. Bieszczad, and B. Pagurek. Application-Oriented Network Modeling with Mobile Agents. In *Proceedings of the IEEE/IFIP Network Operations and Management Symposium (NOMS'98)*, February 1998.

[23] B. Schwartz, A. Jackson, W. Strayer, W. Zhou, R. Rockwell, and C. Partridge. Smart Packets: Applying Active Networks to Network Management. *ACM Transactions on Computer Systems*, 18(1), February 2000.

[24] P. Simões, L. Moura Silva, and F. Boavida-Fernandes. Integrating SNMP into a Mobile Agent Infrastructure. In *Proceedings of the Tenth IFIP/IEEE International Workshop on Distributed Systems: Operations and Management (DSOM'99)*, October 1999.

[25] William Stallings. *SNMP, SNMPv2, SNMPv3, and RMON 1 and 2*. Addison-Wesley, third edition, 1999.

[26] G. Susilo, A. Bieszczad, and B. Pagurek. Infrastructure for Advanced Network Management based on Mobile Code. In *Proceedings of the IEEE/IFIP Network Operations and Management Symposium (NOMS'98)*, February 1998.

[27] T. Thorn. Programming Languages for Mobile Code. *ACM Computing Surveys*, 29(3), September 1997.

# Open Packet Monitoring on FLAME: Safety, Performance, and Applications*

Kostas G. Anagnostakis, Michael Greenwald, Sotiris Ioannidis, and
Stefan Miltchev

CIS Department, University of Pennsylvania
200 S. 33rd Street, Philadelphia PA 19104, USA
{anagnost,mbgreen,si,miltchev}@dsl.cis.upenn.edu

**Abstract.** Packet monitoring arguably needs the flexibility of open architectures and active networking. In earlier work we have implemented
FLAME, an open monitoring system, that balanced flexibility and safety
while attempting to achieve high performance by combining the use of a
type-safe language, lightweight run-time checks, and fine-grained policy
restrictions.

We seek to understand the range of applications, workloads, and traffic,
for which a safe, open, traffic monitoring architecture is practical. To that
end, we investigated a number of applications built on top of FLAME. We
use measurement data and analysis to predict the workload at which our
system cannot keep up with incoming traffic. We report on our experience
with these applications, and make several observations on the current
state of open architecture applications.

## 1 Introduction

The bulk of research on *Active Networks* [23] has been directed towards building
general infrastructure [1, 24], with relatively little research driven by the needs
of particular applications. Recently the focus has shifted slightly as researchers
have begun to investigate issues such as safety, extensibility, performance, and
resource control, from the perspective of *specific* applications [4, 18].

Network traffic monitoring is one such application. [3, 4] makes the case that
network traffic monitoring can benefit greatly from a monitoring infrastructure
with an open architecture, as static implementations of monitoring systems are
unable to keep up with evolving demands. The first big problem is that, in many
cases, monitoring is required at multiple points in the network. No distributed
monitoring infrastructure is currently deployed, so monitoring must typically
take place at the few nodes, such as routers, that already monitor traffic and ex-
port their results. While routers do offer *built-in* monitoring functionality, router
vendors only implement monitoring functions that are cost-effective: those that

---

* This work was supported in part by the DoD University Research Initiative (URI)
program administered by the Office of Naval Research under Grant N00014-01-1-
0795, and by NSF under grant ANI-00-82386.

J. Sterbenz et al. (Eds.): IWAN 2002, LNCS 2546, pp. 120–131, 2002.

are interesting to the vast majority of possible customers. If one needs functions that are not part of the common set, then there may be no way to extract the needed data from the routers. Furthermore, as customer interests evolve, the router vendors can only add monitoring functionality on the time-scale of product design and release; it can be months or years from the time customers first indicate interest until a feature makes it into a product. Therefore, the need for timely deployment cannot always be met at the current pace of standardization or software deployment, especially in cases such as detection and prevention of denial-of-service attacks.

In response to these problems, several prototype extensible monitoring systems [14, 4, 3, 11] have been developed. One basic goal of such approaches is to allow the use of critical system components by users other than the network operator. However, providing users with the ability to run their own modules on nodes distributed throughout the network requires extensible monitoring systems to provide protection mechanisms.

Flexible protection mechanisms, and other methods of enforcing safety, are essential for extensible monitoring systems for two reasons. First, users, such as researchers who want to study network behavior, should not have access to all the data passing through a router. Rather, fine-grained protection is needed to allow the system to enforce policy restrictions, *e.g.*, ensuring privacy by limiting access to IP addresses, header fields, or packet content. Second, protection from interference is needed to guard against poorly implemented (or malicious) modules which could otherwise hurt functions that may be critical to the operation of the network infrastructure.

The thrust of our research is to determine whether programmable traffic monitoring systems that are flexible enough to be useful, and safe enough to be deployed, can perform well enough to be practical.

In LAME [4] we demonstrated that it is possible to build an extensible monitoring system using off-the-shelf components. Further investigation demonstrated performance problems with the use of off-the-shelf components in LAME. Our follow-on project, FLAME, presented a design that preserved the safety properties of LAME, but was designed for high performance. FLAME combines several well-known mechanisms for protection and policy control; in particular, the use of a type-safe language, custom object patches for run-time checks, anonymizing, and namespace protection based on trust management.

The purpose of the study in this paper is to understand the range of applications and traffic rates for which a safe, open, traffic monitoring architecture is practical. In [3] we presented preliminary results that demonstrated that FLAME largely eliminated the performance problems of LAME. We have implemented a number of additional test applications and have used them as our experimental workload. We use the data collected from these applications to quantify and analyze the performance costs, and to predict the workload at which our system will no longer be able to keep up with incoming traffic.

The general tenor of the results reported here (although not the specific numbers) should be more widely applicable than just to FLAME. For example, the

Open Kernel Environment (OKE) of Bos and Samwel [6] adopts a similar approach to FLAME. The OKE designers also carefully considered the interaction between safety features and performance implications. OKE, among other features, provides additional flexibility through the use of trust-controlled elastic language extensions. These extensions provide increased control over the trade-offs between safety and performance, as, for example, certain checks which are hard-wired in our design can be eliminated, if appropriate trust credentials are provided. The work reported in this paper should give some indications about the workload supportable by systems such as OKE, also.

The rest of this paper is structured as follows. A brief overview of the FLAME architecture, including protection mechanisms, is given in Section 2. In Section 3 we study the performance trade-offs of the resulting system, and we conclude in Section 4.

## 2   Overview of the FLAME Architecture

The architecture of FLAME is shown in Figure 1. A more detailed description is available in [3]. Modules consist of kernel-level code $K_x$, user-level code $U_x$, and a set of credentials $C_x$. Module code is written in Cyclone [12] and is processed by a trusted compiler upon installation. The kernel-level code takes care of time-critical packet processing, while the user-level code provides additional functionality at a lower time scale and priority. This is needed so applications can communicate with the user or a management system (*e.g.,* using the standard library, sockets, *etc.*).

There has been a small architectural modification to FLAME since the publication of [3], after experimentation under high load. The original FLAME architecture interacted with the network interface exclusively through interrupts. As others have noted [15, 22], under high rates of incoming network traffic, interrupt handling can degrade performance. More recent versions of FLAME poll the network interface card (NIC) to read packets to avoid performance degradation. Note that the polling technique and the resulting performance improvement is well known and does not represent a contribution of this paper.

In terms of deployment, the system can be used as a passive monitor *e.g.* by tapping on a network link by means of an optical splitter, or using *port mirroring* features on modern switches. Ideally, a FLAME -like subsystem would be part of an enhanced router interface card. A preliminary study shows how such a subsystem can be built using a network processor board [2]. For the purposes of this paper, we consider FLAME in a passive monitor set-up.

The basic approach is to use the set of credentials, $C_x$, at compile time to verify that the module is allowed by system policy to perform the functions it requests. The dark units in Figure 1 beside each $K_x$ represent code that is inserted before each module code segment for enforcing policy-related restrictions. These units appropriately restrict access of modules to packets or packet fields, provide selective anonymization of fields, and so on.

For allowing user code to safely execute inside the operating system kernel, the system needs to guard against excessive execution time, privileged instruc-

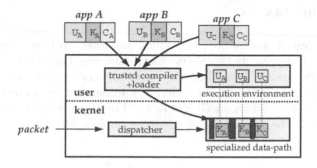

**Fig. 1. FLAME Architecture**

tions, exceptions and random memory references. There has been extensive work in the operating system and language communities that addresses the above problems (c.f. [20, 8, 25]). FLAME leverages these techniques to satisfy our security needs.

**Bounding Execution Time.** For bounding execution time we take an approach similar to [10]: we augment the backward jumps with checks to a cycle counter; if the module exceeds its allocated execution time we jump to the next module. On the next invocation, the module can consult an appropriately set environment variable to check if it needs to clean-up data or exit with an error. This method adds an overhead of 5 assembly instructions for the check. If the check succeeds there is an additional overhead of 6 instructions to initiate the jump to the next module.

**Exceptions.** We modified the trap handler of the operating system to catch exceptions originating from the loaded code. Instead of causing a system panic we terminate the module and continue with the following one.

**Privileged Instructions and Random Memory References.** We use Cyclone [12] to guard against instructions that may arbitrarily access memory locations or may try to execute privileged machine instructions. Cyclone is a language for C programmers who want to write secure, robust programs. It is a dialect of C designed to be *safe*: free of crashes, buffer overflows, format string attacks, and so on. All Cyclone programs must pass a combination of compile-time, link-time and run-time checks to ensure safety.

**Policy control.** Before installing a module in our system we perform policy compliance checks[1] on the credentials this module carries. The checks determine the privileges and permissions of the module. In this way, the network operator is able to control what packets a module can access, what part of the packet a module is allowed to view and in what way, what amount of resources (processing, memory, *etc.*) the module is allowed to consume on the monitoring system, and what other functions (*e.g.*, socket access) the module is allowed to perform.

---

[1] Our policy compliance checker uses the KeyNote [5] system.

# 3   Experiments

This section describes a number of applications that we have implemented on FLAME and then presents three sets of experiments. The first involves the deployment of the system in a laboratory testbed, serving as a proof of concept. The second looks at issues of the underlying infrastructure, in order to specify the *capacity of our system* on Gbit/s links. The third set of experiments provides a picture of the processing cost of our example applications, and protection overheads.

## 3.1   Applications

We present examples of applications that a) are widely regarded as useful but appear to be stalled in the standardization process (trajectory sampling), b) would be difficult to deploy in time to be useful (worm detection) and c) may be valuable in certain situations but may not be globally useful to make it worth implementing in routers (RTT analysis, LRD analysis).

*Trajectory sampling.* Trajectory sampling, developed by Duffield and Grossglauser [9], is a technique for coordinated sampling of traffic across multiple measurement points, effectively providing information on the spatial flow of traffic through a network. The key idea is to sample packets based on a hash function over the invariant packet content (*e.g.* excluding fields such as the TTL value that change from hop to hop) so that the same packet will be sampled on all measured links. Network operators can use this technique to measure traffic load, traffic mix, one-way delay and delay variation between ingress and egress points, yielding important information for traffic engineering and other network management functions. Although the technique is simple to implement, we are not aware of any monitoring system or router implementing it at this time.

We have implemented trajectory sampling as a FLAME module that works as follows. First, we compute a hash function $h(x) = \phi(x) \mod A$ on the invariant part $\phi(x)$ of the packet. If $h(x) > B$ , where $B < A$ controls the sampling rate, the packet is not processed further. If $h(x) < B$ we compute a second hash function $g(x)$ on the packet header that, with high probability, uniquely identifies a flow with a label (*e.g.* TCP sequence numbers are ignored at this stage). If this is a new flow, we create an entry into a hash table, storing flow information (such as IP address, protocol, port numbers *etc.*). Additionally, we store a timestamp along with $h(x)$ into a separate data structure. If the flow already exists, we do not need to store all the information on the flow, so we just log the packet.

*Round-trip time analysis.* We have implemented a simple application for obtaining an approximation of round-trip delays for TCP connections passing through a link. The round-trip delay is an important metric for understanding end-to-end performance due to its role in TCP congestion control [13]. Additionally, measuring the round-trip times observed over a specific ISP provides a reasonable indication of the quality of the service provider's infrastructure, as well as its connectivity to the rest of the Internet. Finally, observing the evolution of round-trip delays over time can be used to detect network anomalies on shorter time scales, or to observe the variation in service quality over longer periods of time.

The implementation is both simple and efficient. We watch for TCP SYN packets indicating a new connection request, and watch for the matching TCP ACK packet (in the same direction). The difference in time between the two events provides a reasonable approximation of the round-trip time between the two ends of the connection. For every SYN packet received, we store a timestamp into a hash-table. As the first ACK after a SYN usually has a sequence number which is the SYN packet's sequence number plus one, this number is used as the key for hashing. Thus, in addition to watching for SYN packets, the application only needs to look into the hash table for every ACK received. The hashtable can be appropriately sized depending on the number of flows and the desired level of accuracy.

*Worm detection.* The concept of "worms" and techniques to implement them have existed since the early descriptions in [7, 21]. A worm compromises a system such as a Web server by exploiting system security vulnerabilities; once a system has been compromised the worm attempts to replicate by "infecting" other hosts. Recently, the Internet has observed a wave of "worm" attacks [16]. The "Code Red" worm and its variants infected over 300,000 servers in July-August 2001.

This attack can be locally detected and prevented if the packet monitor can obtain access to the TCP packet content. Unfortunately, most known packet monitors only record the IP and TCP header and not the packet payload. We have implemented a module to scan packets for the signature of one strain of "Code Red" (the random seed variant). If this signature is matched, the source and destination IP addresses are recorded and can be used to take further action (such as blocking traffic from attacking or attacked hosts *etc.*). Despite the ability to locally detect and protect against worms, widespread deployment of an extensible system such as FLAME would still have improved the fight against the virus.

*Real-time estimation of long-range dependence parameters.* Roughan *et al.* [19] proposed an efficient algorithm for estimating long-range dependence parameters of network traffic in real-time. These parameters directly capture the variability of network traffic and can be used, beyond research, for purposes such as measuring differences in variability between different traffic classes and characterizing service quality. We have ported the algorithm to Cyclone and implemented the appropriate changes to allow execution as a module on the FLAME system. Some modifications were needed for satisfying Cyclone's static type checker and providing appropriate input, *e.g.,* traffic rates over an interval. The primary difference between this module and the other applications is that separate kernel and user space components were needed. This requirement arises because the algorithm involves two loops: the inner loop performs lightweight processing over a number of samples, while the outer loop performs the more computationally intensive task of taking the results and producing the estimate. As the system

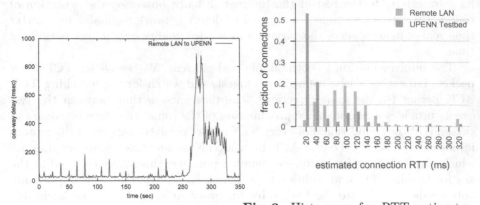

**Fig. 3.** Histogram for RTT estimates for the same targets seen from two networks.

**Fig. 2.** Measuring one way delay between two networks.

cannot interrupt the kernel module and provide scheduling, the outer loop had to be moved to user space.

### 3.2    Experiment Setup

The testbed used for our experiments involves two sites: a local test network at Penn, and a remote LAN connecting to the Internet through a DSL link. The minimum round-trip delay between the two sites is 24 ms. The test network at Penn consists of 4 PCs connected to an Extreme Networks Summit 1i switch. The switch provides port mirroring to allow any of its links to be monitored by the FLAME system on one of the PCs. All PCs are 1 GHz Intel Pentium III with 512 MB memory, OpenBSD 2.9 operating system except for the monitoring capacity experiments where we used the Click [17] code under Linux 2.2.14 on the sending host. The FLAME system uses the Intel PRO/1000SC Gigabit NIC.

### 3.3    Testbed Demonstration

In this section we demonstrate the use of the round-trip delay analysis and trajectory sampling modules on our experimental setup. We have installed the round-trip delay analysis module on the two FLAME monitors, on the remote LAN and the PENN test network. We initiated wget to recursively fetch pages, starting from the University of Pennsylvania main web server. In this way we created traffic to a large number of sites reachable through links on the starting Web page. The experiment was started concurrently on both networks to allow us to compare the results. One particular view of 5374 connections over a 1 hour period is presented in Figure 3, clearly showing the difference in performance which is partly due to the large number of local or otherwise well connected sites that are linked through the University's Web pages.

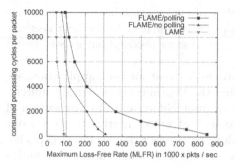

Fig. 4. Available processing cycles per packet as a function of maximum loss-free traffic rate.

| Module | gcc | Cyclone | Cyclone protection | Cyclone protection optimized |
|--------|-----|---------|--------------------|------------------------------|
| Traj.smpl. | 381 | 10.2% | 20.2% | 12.8% |
| RTT est. | 183 | 12.4% | 15.3% | 15.3% |
| Worm det. | 24 | 83.3% | 125% | 83.3% |
| LRD est. | 143 | 7.6% | 10.4% | 9% |

Fig. 5. Module processing costs (in cycles), Cyclone overhead, FLAME protection overhead, and optimization effect.

We also executed the trajectory sampling module and processed the data collected by the module to measure the one way delay for packets flowing between the two networks. The clocks at the two monitors were synchronized using NTP prior to the experiment. The results are shown in Figure 2. Note that this is different from simply using `ping` to sample delays, as we measure the *actual* delay experienced by network traffic. The spike shows our attempt to overload the remote LAN using UDP traffic.

### 3.4   System Performance, Workload Analysis, and Safety Overheads

We determine how many processing cycles are available for executing monitoring applications at different traffic rates. We report on the performance of FLAME with and without the interface polling enhancement as well as LAME.

The experiment is designed as follows. Two sender PCs generate traffic to one sink, with the switch configured to mirror the sink port to the FLAME monitor. The device driver on the FLAME system is modified to disable interrupts and the FLAME system is instrumented to use polling for reading packets off the NIC. To generate traffic at different rates, we use the Click modular router system under Linux on the sending side. All experiments involve 64 byte UDP packets. The numbers are determined by inserting a busy loop into a null monitoring module consuming processing cycles. The sending rate is adapted downward until no packets are dropped at the monitor. This may seem overly conservative, because packet losses occur when even one packet is delivered to FLAME too early. However, the device driver allocates 256 RxDescriptors for the card to store 2K packets. Therefore the card can buffer short-term bursts that exceed the average rate without incurring packet loss, but cannot tolerate sustained rates above the limit. In Figure 4 we show the number of processing cycles available at different traffic rates, for LAME, FLAME without polling, and FLAME with polling enabled.

There are two main observations to make on these results. First, as expected, the polling system performs significantly better, roughly 2.5 times better than the non-polling system. Second, the number of cycles available for applications to consume, even at high packet rates, appears reasonable. In the next sections we will discuss these figures in light of the processing needs of our experimental applications.

To obtain an rough estimate of the processing cost for each application, we instrumented the application modules using the Pentium performance counters. We read the value of the Pentium cycle counter before and after execution of application code for each packet. Due to lack of representative traffic on our laboratory testbed, we fed the system with packets using the Auckland-II packet trace provided by NLANR and the WAND research group. The measurements were taken on a 1 GHz Intel Pentium III with 512 MB memory, OpenBSD 2.9 operating system, gcc version 2.95.3, and Cyclone version 0.1.2.

We compare the processing cost of a pure C version of each application to the Cyclone version, with and without protection, and using additional optimizations to remove or thin the frequency of backward jumps (these modifications were done by hand). We measure the median execution time of each module over 113 runs. The results from this experiment are summarized in Table 5.

There are four main observations to make. First, the cost per-application appears to be well within the capabilities of a modern host processor, for a reasonable spectrum of traffic rates. Second, the cost of protection (after optimization), does not exceed by far the cost of an unprotected system. Third, the costs presented are highly application dependent and may therefore vary. Finally, some effort was spent in increasing the efficiency of both the original C code as well as the Cyclone version Thus, care must be taken not to overstate these results. This experiment *does* indicate that it is feasible to provide protection mechanisms in an open monitoring architecture, enabling support for experimental applications and untrusted users. However, the numbers should not be considered representative of off-the-shelf compilers and/or carelessly designed applications.

## 3.5   Modeling Supportable Workloads and Traffic Rates

We can roughly model the expected performance (maximum supportable packet rate) of FLAME as a function of workload (number of active modules). We derive the model from our measured system performance from Section 3.4, and the costs of our experimental applications and the measured safety overheads from Section 3.4.

We can approximately fit the number of available cycles to $a_0 r^{b_0}$, where $r$ is transmission rate in packets per second and $a_0$ and $b_0$ are constants. Computing $a_0$ and $b_0$ using least squares, and dropping the data point at 848k packets per second[2], we get the number of available cycles for processing is $3 \times 10^9 r^{-1.1216}$.

---

[2] The fit is remarkably good for packet rates under 500,000 packets per second. The fit is good for packet rates up to about 800,000 packets per second, but our measurements when the gigabit network was running full bore sending 64 byte packets (small), yielded fewer available cycles than predicted by our model.

Packets per second, $r$, can itself be computed as $B/8s$ where $B$ is the transmission rate in bits per second, and $s$ is the mean packet size in bytes. Assuming a mean module computation cost of 210 cycles per module (based on the assumption that our applications are representative), and using our measured overhead of 60 cycles per module, we can support a workload of $\lfloor \frac{1}{9} 10^8 r^{-1.1216} \rfloor$ modules for an incoming traffic rate of $r$ packets per second, without losing a single packet. Conversely, we can compute the maximum traffic rate as a function of the number of available cycles, $c$, by $r = 2.816 \times 10^8 c^{-0.8916}$ (or $r = 1.914 \times 10^6 n^{-0.8916}$, where $n$ is the number of modules).

To apply this model on an example, consider a fully-utilized 1 Gbit/s link, with a median packet size of 250 bytes, which is currently typical for the Internet. In this scenario, $r$, the input packet rate, is approximately 500,000 packets per second. The model predicts enough capacity to run 5 modules. For comparison, note that we measured the maximum loss-free transmission rate for 1310 cycles on a 1 Ghz Pentium to be 500,004 packets per second; 1310 cycles comfortably exceeds the total processing budget needed by the 4 applications in this study (841 cycles with safety checks, and 731 cycles without any safety checks). Alternatively, with 20 active modules loaded, and an average packet size of 1K bytes (full-size ethernet data packets, with an ack every 2 packets), the system can support a traffic rate over 1 Gbps.

The demonstrated processing budget may appear somewhat constrained, assuming that users may require a much richer set of monitoring applications to be executed on the system. However, in evaluating the above processing budget, three important facts need to be considered. First, faster processors than the 1 GHz Pentium used for our measurements already exist, and processors are likely to continue improving in speed. Second, a flexible system like FLAME may not be required to cover *all* monitoring needs: one can assume that some portion of applications will be satisfied by static hardware implementation in routers, with an open architecture supporting only those functions that are not covered by the static design. Third, the figures given above represent the rate and workload at which no packets are lost. As the number of active applications increases, it will be worthwhile to allow the system to degrade gracefully. The cost of graceful degradation is an increase in the constant per-module overhead due to the added complexity of the scheduler — thus packet loss will occur under slightly lighter load than in the current configuration, but an overloaded system will shed load gracefully.

Based on our results, we can assert that FLAME is able to support a reasonable application workload on fully loaded Gbit/s links. Using FLAME on higher network speeds (*e.g.* 10 Gbit/s and more) does not currently seem practical and is outside the scope of our work.

## 4   Summary and Concluding Remarks

We have spent some time building, measuring, and refining an open architecture for network traffic monitoring. Several interesting observations are worth reporting:

*The techniques developed to build general infrastructure are applicable and portable to specific applications.* LAME was built using off-the-shelf components. FLAME, in contrast, required us to write custom code. However, it was constructed using "off-the-shelf technology". That is, the techniques we used for extensibility, safety, and efficiency were well-known, and had already been developed to solve the same problems in a general active-networking infrastructure. In particular, the techniques used for open architectures are now sufficiently mature that applications can be built by importing technology, rather than by solving daunting new problems.

Nevertheless, *careful design is still necessary.* Although the technology was readily available, our system has gone through three architectural revisions, after discovering that each version had some particular performance problems. Care must be taken to port the *right* techniques and structure, otherwise the price in performance paid for extensibility and safety may render the application impractical.

Programmable applications are clearly more flexible than their static, closed, counterparts. However, to the limited extent that we have been able to find existing custom applications supporting similar functionality, we found that *careful engineering can make applications with open architectures perform competitively with custom-built, static implementations.*

More experience building applications is certainly needed to support our observations, but our experience so far supports the fact that high performance open architecture applications are practical.

# References

[1] D. S. Alexander, W. A. Arbaugh, M. W. Hicks, P. Kakkar, A. D. Keromytis, J. T. Moore, C. A. Gunter, S. M. Nettles, and J. M. Smith. The SwitchWare active network architecture. *IEEE Network*, 12(3):29–36, May/June 1998.

[2] K. G. Anagnostakis and H. Bos. Towards flexible real-time network monitoring using a network processor. In *Proceedings of the 3rd USENIX/NLUUG SANE Conference (short paper)*, May 2002.

[3] K. G. Anagnostakis, S. Ioannidis, S. Miltchev, J. Ioannidis, M. B. Greenwald, and J. M. Smith. Efficient packet monitoring for network management. In *Proceedings of the 8th IFIP/IEEE Network Operations and Management Symposium (NOMS)*, pages 423–436, April 2002.

[4] K. G. Anagnostakis, S. Ioannidis, S. Miltchev, and J. M. Smith. Practical network applications on a lightweight active management environment. In *Proceedings of the 3rd Int'l Working Conference on Active Networks (IWAN)*, pages 101–115, October 2001.

[5] M. Blaze, J. Feigenbaum, J. Ioannidis, and A. D. Keromytis. The KeyNote Trust Management System Version 2. Internet RFC 2704, September 1999.

[6] H. Bos and B. Samwel. Safe kernel programming in the OKE. In *Proceedings of IEEE OPENARCH 2002*, June 2002.

[7] J. Brunner. *The Shockwave Rider*. Del Rey Books, Canada, 1975.

[8] J. Chase, H. Levy, M. Baker-Harvey, and E. Lazowska. Opal: A single address space system for 64-bit architectures. In *Proceedings of the Fourth Workshop on Workstation Operating Systems*, pages 80–85, 1993.

[9] N. Duffield and M. Grossglauser. Trajectory sampling for direct traffic observation. *IEEE/ACM Transactions on Networking*, 9(3):280–292, June 2001.

[10] M. Hicks, J. T. Moore, and S. Nettles. Compiling PLAN to SNAP. In *Proceedings of the 3rd Int'l Working Conference on Active Networks (IWAN)*, pages 134–151, October 2001.

[11] S. Ioannidis, K. G. Anagnostakis, J. Ioannidis, and A. D. Keromytis. xPF: packet filtering for low-cost network monitoring. In *Proceedings of the IEEE Workshop on High-Performance Switching and Routing (HPSR)*, pages 121–126, May 2002.

[12] T. Jim, G. Morrisett, D. Grossman, M. Hicks, J. Cheney, and Y. Wang. Cyclone: A safe dialect of C. In *Proceedings of USENIX 2002 Annual Technical Conference*, June 2002.

[13] T. V. Lakshman and U. Madhow. The performance of TCP/IP for networks with high bandwidth-delay products and random loss. *IEEE/ACM Transactions on Networking*, 5(3):336 – 350, June 1997.

[14] G. R. Malan and F. Jahanian. An extensible probe architecture for network protocol performance measurement. In *Proceedings of ACM SIGCOMM*, pages 215–227, August 1998.

[15] J. C. Mogul and K. K. Ramakrishnan. Eliminating receive livelock in an interrupt-driven kernel. *ACM Transactions on Computer Systems*, 15(3):217–252, August 1997.

[16] D. Moore. The spread of the code-red worm (crv2). In *http://www.caida.org/analysis/security/code-red/*. August 2001.

[17] R. Morris, E. Kohler, J. Jannotti, and M. F. Kaashoek. The click modular router. In *Proceedings of the 17th ACM Symposium on Operating System Principles (SOSP)*, pages 217–231, December 1999.

[18] C. Partridge, A. Snoeren, T. Strayer, B. Schwartz, M. Condell, and I. Castineyra. FIRE: Flexible intra-AS routing environment. In *Proceedings of ACM SIGCOMM*, pages 191–203. August 2000.

[19] M. Roughan, D. Veitch, and P. Abry. Real-time estimation of the parameters of long-range dependence. *IEEE/ACM Transactions on Networking*, 8(4):467–478, August 2000.

[20] F. B. Schneider, G. Morrisett, and R. Harper. A language-based approach to security. *Informatics: 10 Years Back, 10 Years Ahead*, pages 86–101, 2000.

[21] J. F. Shoch and J. A. Hupp. The "worm" programs – early experiments with a distributed computation. *Communications of the ACM*, 25(3):172–180, March 1982.

[22] J. M. Smith and C. B. S. Traw. Giving applications access to Gb/s networking. *IEEE Network*, 7(4):44–52, July 1993.

[23] D. Tennenhouse, J. Smith, W. Sincoskie, D. Wetherall, and G. Minden. A survey of active network research. *IEEE Communications Magazine*, pages 80 – 86, January 1997.

[24] D. Wetherall. Active network vision and reality: Lessons from a capsule-based system. In *Proceedings of the 17th ACM Symposium on Operating System Principles (SOSP)*, pages 64 – 79, December 1999.

[25] C. Yarvin, R. Bukowski, and T. Anderson. Anonymous RPC: Low-latency protection in a 64-bit address space. In *Proceedings of the 1993 Summer USENIX Conference*, June 1993.

# Active Networks for 4G Mobile Communication: Motivation, Architecture and Application Scenarios

Christian Prehofer, Qing Wei

DoCoMo Communications Laboratoires Europe,
Landsberger Str. 338-362, Munich, Germany
{prehofer, wei}@docomolab-euro.com

**Abstract.** In this paper, we examine the application of active networking technology to future mobile networks. We first introduce an architecture for programmable 4th generation (4G) mobile networking, including all system layers on the network and terminal side. Based on this architecture, we discuss programmability in future mobile networks. We investigate the main driving forces and obstacles for the application of active networks. In particular, we show that flexible component installation and cross layer interfaces are a main motivation for programmable mobile networks. This is illustrated by a number of applications for future mobile networks, including context aware mobility management and paging, where flexibility is a key requirement for future mobile services.

## 1   Introduction

In this paper, we present the driving forces for applying active network technology to future mobile networks. Active networks introduce flexibility into network elements by adding programmable execution environments for deploying new software. For the successful adaptation of active networks in future mobile networks, we think that there has to be a clear common vision. For this purpose, we first introduce an architecture for programmable 4th generation (4G) mobile networking, addressing networking on both terminals and network elements. Based on this, we discuss requirements regarding flexibility and programmability for 4G.

In our view, a main impediment for the usage of active networks is that active networks are, to some degree, a disruptive technology [1]. Active networking forms a significant change in the software architecture of commercial network elements. In particular, active networks are likely to be more expensive in the beginning. This is mainly due to the overhead of execution environment regarding performance, memory usage and code deployment infrastructure. We hence examine the main driving forces and obstacles for the application of active networks. In particular, we show that flexible component installation and cross layer interfaces are a main motivation for programmable networks. In order to enable this programmability, active networking has to support reconfiguration and changing interfaces in different layers.

In the following, we first introduce active networking and examine different classes of programmability. Then we present in Section 2 an architecture for mobile

J. Sterbenz et al. (Eds.): IWAN 2002, LNCS 2546, pp. 132–145, 2002.

networks with the focus on programmable technologies. This shows that programmability is a pervasive concept in future mobile networks. In Section we discuss the driving forces for programmability with the special focus on mobile networks. This is illustrated in Section by a number of application areas for future mobile networks including context aware mobility management.

## Active Networking and Programmability

Active networking is a promising technology in which the network nodes are programmed to perform customized operations on the packets that pass through them. A programmable node will enable the fast deployment of new services by using open interfaces to the network resources. With this technology new protocols can be installed on the network elements which use the lower layer resources for creating new services. Active network technology typically provides an execution environment which is operating system and hardware independent. There are several approaches on active networks which include AMNet [ ], ANTs [ ], ANN [ ], CANEs [ ], Mobiware [ ]. They focus on different architectural domains for instance quality of service (QoS) control, signalling, management, transport, and applications. Some approaches aim for high performance needed for flexible per packet processing while others only aim at flexibility in the control path.

Another related technology for programmability is software defined radio in reconfigurable terminals [ ] or base stations. This technology will add flexibility to the lower layers including RF, base band and networking protocols. For instance terminals will not only have platforms for installing new applications but also the radio hardware and the stacks will be programmable.

To classify these technologies we consider the following different classes of programmability.

- Parameterization of software or hardware modules is the classical way to introduce flexibility without the need for software updates.
- Complete SW update or exchange. This is typically a firmware update of a device, e.g. a mobile phone or hardware module software update. This kind of update often requires that the device has to be disabled for some time period which is clearly not desirable.
- Partial SW update of a component or a software module in a complex software system. In this case there are no open interfaces and the correct functioning of the complete software has to be reconsidered. The update may also lead to service interruptions.
- Installation of a software module in an open preferably virtual execution environment. In this case the environment and the interface design should ensure proper functionality if the component behaves properly. In this case a software update should not lead to service disruptions.
- In the capsule approach e.g. [ ] the program code is contained in packets. This approach is possible in some active networks environments where the packets include the code needed to handle these packets.

For active networks we consider typically the classes above while software radio [ ] also employs update of classes for internal hardware modules. Currently network equipment and mobile terminals are mostly capable of class and updates, and may in future be capable of some class or updates.

In the following section we present an abstract model for mobile network elements and devices. While this model presents a very homogenous view of the different layers the platforms on these layers may use different classes of programmability.

## G Programmable Mobile Network Architecture

In this section we introduce a high level architecture for future mobile systems with the focus on programmability. To address the creation and provisioning of unanticipated services the whole system has to be designed to be as flexible as possible. Openness and configurability not only at the service level but also with respect to the whole system architecture will invite third party vendors to evolve the system as it unfolds and is therefore the key to viable solutions.

There exist many visions on fourth generation mobile networks G [ ] and the WWRF [ ] is currently working on a combined vision and research agenda. G systems are expected to provide an integration platform to seamlessly combine heterogeneous environments (e.g. see [ ]). Core abstraction layers cover hardware platform, network platform, middleware platform, and applications

The abstraction layers can interact with each other using well defined interfaces. Besides their regular cooperation in an operational setting, each layer can be configured separately and independently via configuration interfaces.

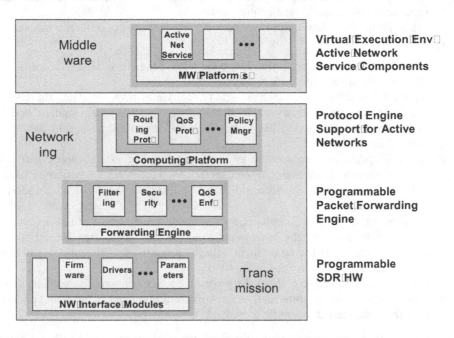

**Fig Programmable Network Element Model**

## Network Element Architecture

Our generic architecture for the network elements of a mobile network is shown in **Fig** 2 excluding applications. In this architecture we consider the following abstraction layers each programmable with configurable components:

- Middleware platform typically with a virtual execution environment for platform independent distributed processing.
- The computing platform serves as a general purpose platform for processing stateful protocols (e.g. routing, QoS signaling or connection management).
- The forwarding engine is in the data path of a network node and it connects the interface modules (e.g. by a switch matrix). This engine can be implemented as dedicated hardware or as a kernel module of common operating systems. The forwarding engine is programmable for performance critical tasks which are performed on a per packet basis.
- The interface modules are medium specific for different wireless or wired standards. They can be configured or programmed to adapt to new physical layer protocols or for triggering events in higher layers.

It is instructive to discuss the different approaches to active networking in this architecture. Some approaches offer a virtual execution environment (e.g. a Java virtual machine) on the middleware layer. Some approaches also include native code in the computing platform (e.g. for flexible signaling). Others employ programmable hardware for forwarding [11]. A key ingredient in these approaches is interfaces to the lower layers and programmable filters for identifying the packets to be processed in the active networking environment.

## Mobile Terminal Architecture

In the following we discuss an architecture for mobile terminals. Although active networking traditionally refers to network elements and less on terminals we think that the terminals have to be included as well for end to end services. Furthermore, with upcoming ad hoc or multi hop networks the conceptual distinction between network elements is blurred. The main components of the terminal architecture shown in **Fig** 3 are:

- Middleware platform typically with a limited virtual execution environment.
- Smart Card (e.g. USIM for UMTS) which includes subscriber identities and also a small but highly secure execution environment. This can be used ideally for personal services like electronic wallet.
- Programmable hardware which is designed for one or more radio standards.
- Native operating system which provides real time support needed for stacks and certain critical applications (e.g. multimedia codecs).

Compared to network elements the SmartCard is a new programmable component. Due to resource limitations the Forwarding Engine and Computation Platform just collapse to one operating system platform. Also the middleware layers are typically quite restricted.

Service deployment and control of reconfigurations are complex since there is a split of responsibility between operator and manufacturer. For instance, the manufacturer has to provide or digitally sign appropriate low level code for reconfigurations. On the other hand, the operator is interested in controlling the configuration to fit the user and network needs.

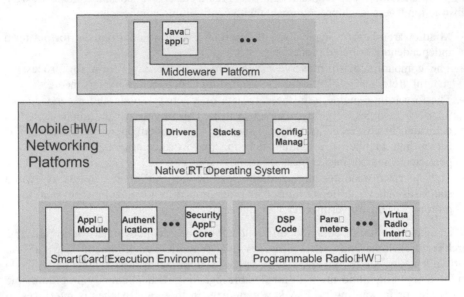

**Fig. Mobile Terminal Architecture**

## Driving Forces for Active Network Technology

Active networking technology is a major step towards flexibility in networking. This follows a long term technology trend towards open, virtual execution platforms in many application areas. While there has been extensive research on active networks, their commercial application is still in its infancy. The vision of a fully programmable network is difficult to achieve since one has to balance the additional costs for active networks with their benefits. The main cost factors are more expensive hardware, development of a reliable active networking platform, integration with existing software etc. The main benefits are increased flexibility for new software and more efficient software development, including maintenance.

Since this increased flexibility is the main stronghold of active networks, we focus in the following on this. Typically, one assumes that more flexibility is needed on the application and middleware layers or higher, and less on the networking layer. For mobile networks, we argue that flexibility in networking is needed as well for the following reasons.

- Mobile networks are very fragile, due to wireless links and mobility
- Mobile networks are more expensive than wired networks, hence optimizations pay off more easily
- Considerable innovation in air interfaces will require adaptation for new technologies

We show in the following that flexible component installation and cross layer interfaces are a key ingredient for wireless network and can be ideally supported by active networks.

## Cross Layer Interfaces and Active Networks

Networking protocols are traditionally classified in layers based on the ISO OSI classification of layers. This is an important abstraction needed for modularity and for managing the complexity of current networking systems. However in mobile networking one often needs specific and possibly medium dependent interfaces between the layers as e.g. discussed in []. Cross layer interfaces define APIs between modules in different layers which exchange information beyond the standard interfaces between the layers.

In the classical view cross layer interfaces may break modularity and the important concept of abstraction. The main benefit of this abstraction is that layers can be exchanged modularly, e.g. a different air interface should not impact the higher layers. However in wireless networks the layer and above often need direct interfaces to the layer e.g. for hand over support. More examples are discussed in [][]. These cross layer interfaces can be classified as follows.

− Additional information about the network e.g. bandwidth or delay variations
− Notification of imminent changes in the network services e.g. hand over or QoS degradation
− Notification of actual changes in the network service
− Configuration APIs e.g. setting parameters for wireless interfaces and also the updating of the components reconfiguration.
− Information from the applications including user context

The problem is that standardizing cross layer interfaces is inherently difficult. Furthermore once standards have been established these interfaces have to be supported once and for all even if they are outdated or not suitable any more. In summary cross layer interfaces are difficult to maintain because cross layer interfaces are often media or application specific and evolve fast.

We claim that active networks are an important technology to install components with cross layer interfaces. Programmable networks can resolve the above problems by introducing "dynamic" cross layer interfaces. Instead of static built in cross layer interfaces the components providing the interfaces are installed when needed.

By dynamically installing APIs we can avoid many problems with static interfaces as described above. This concept is ideal for medium specific API since the components can be installed in a medium specific way. For instance different APIs can be installed for each medium depending on the desired services. This notion of service deployment architecture has been investigated for active networks in [] but not in the context of cross layer interface.

From this discussion we argue that active networking and programmability on other layers has to be integrated. Lower layer reconfiguration e.g. software radio may not use an open execution platform but often use simpler forms of software deployment. However this programmability has to be combined with active networking to offer a full solution.

## Application Examples

In this section we will sketch a number of examples to show the importance of flexible networking protocols and the need of dynamic cross layer interfaces in future mobile networks. Mobility management and hand over optimization are central services of mobile networks. Theses services can often be optimized in many ways, for instance by using the context information available on other layers. Furthermore, applications can be optimized with mobility management. For instance flexible quality of service and adaptation of services is a common example of active networking, as discussed below.

### Context aware Hand Over

In this example we show that hand over can be optimized by information about the user context. As an example, we consider the optimal selection of a new access point during hand over, as shown in **Fig** . In this scenario the terminal moves into an area covered by both AP  and AP . The problem is to decide which Access Point (AP) to choose. State of the art are many algorithms based on signal strength analysis or on available radio resources. Even if one AP is slightly better regarding these local measurements, the decision may not be the best. For instance if the terminal in **Fig** is on the train, it is obviously better to hand over to AP , even if AP  is better reachable for a short period of time.

In many cases, the hand over can be optimized by knowledge of terminal movement and user preferences. For instance if the terminal is in a car or train, its route may be constrained to certain areas. Also, the terminal profile may contain the information that the terminal is built in a car. Alternatively, the movement pattern of a terminal may suggest that the user is in a train.

A main problem is that hand over decisions have to be executed fast. However, the terminal profile and location information is often available on a central server in the core network. Retrieving this information may be too slow for hand over decisions. Furthermore, the radio conditions during hand over may be poor and hence limit such information exchange.

The idea of the solution is to proactively deploy a context aware decision algorithm onto the terminal which can be used to assist hand over decisions. A typical example is the information about the current movement pattern, e.g. by knowledge of train or road routes. For implementation, different algorithms can be deployed by the network on the terminal, depending on the context information. This implementation needs a cross layer interface, which collects the context information from different layers and makes an optimized decision about deploying a decision algorithm.

Similar ideas have already been investigated in [  ] where a middleware was presented which provided a hook for a function which determines the best next access point. In [  ] algorithms based on a static user profile are presented which select among different networks for hand over.

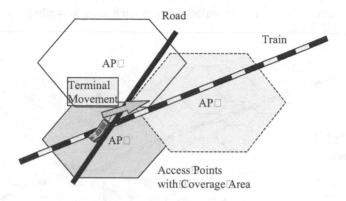

Fig. 4. Context Aware Hand Over Prediction

## 5 Customized Paging

In mobile networks, the network needs to know the location of the mobile terminal (MT) in order to keep the connection with it. This requires the mobile terminal to send the network a location update continuously. A fast moving MT has to send the location update frequently, which causes considerable signaling overhead. For this reason, paging is used instead to save the energy and decrease the signal overhead of location update when a MT is in idle state [11].

With paging, the complex location registration to the networks needs to be performed only when the idle MT moves to another paging area. One paging area consists of several cells with access points. In order to receive an incoming call, the paging process is used to find the exact location of MT in the paging area. The paging strategy can be "Blanket Polling" or "Sequential Paging". In the first strategy, a paging request is sent to all the wireless APs in the paging area simultaneously, then the AP in the cell where the MT located reply the paging request. In the second strategy, paging requests are sent to the APs sequentially in decreasing order of the likelyhood that the MT is located in that cell.

The size and shape of the paging area are essential for the paging efficiency. If large paging areas are chosen, the cost of the paging process will increase, otherwise the rate of registrations and battery consumption will increase. In current systems, a fixed paging area size is used. It is of course not optimal. Active network technology facilitates the customization of the paging area, thus improving the paging performance. Similar to the customized handover, a customized paging also employs the user profile and mobility information to dynamically adjust the paging area.

Fig. 5 gives an example of customized paging area. Suppose the MT is on the train, then the direction and speed of its movement can be determined. In this case, the paging area outside the train route is not necessary. The optimized paging area should be along the railway track as shown in Fig. 5. Because normally trains move very fast, a bigger paging size is preferred to avoid the frequent location registration. Otherwise, if the MT's movement is slow and unpredictable, e.g. pedestrian's mobile phone, the optimal paging area should be centered around the MT and has smaller size, because the registration cost is low. From the above examples, we can see that it will be more efficient to use an adaptive paging area according to the mobility information of the MT.

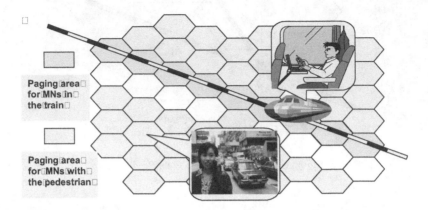

**Fig. 3.** Example of Different Paging Areas for Different MTs

Using active network technology, a program for the paging area calculation can be loaded to the paging area control node according to different scenarios, e.g. the algorithm and input parameters can be changed. The program can be loaded to the computing platform or execution environment in the networking platform. The local context of the MT decides which computation program to load. For example, the user profile may be obtained from the smart card.

There is some literature on the optimization of paging areas. For instance, [ ] estimates the movement of the MT by sending samples. Parameters such as speed, direction are computed based on the samples received. Then the paging area is calculated from the movement parameters. In [ ], a behavior based strategy (BBS) is proposed to estimate the mobile's location by collecting the mobile's long term moving logs. A similar approach is discussed in [ ]. With the active networks solution we have the flexibility to load different algorithms. The input parameters can be changed dynamically according to different user profiles. Obviously, interaction between network layer and the application layer are needed in this application as well.

In the following, we use blanket polling as an example to describe an implementation of customized paging area using our architecture, as shown in **Fig. 4**.

- The network side selects blanket polling as the paging protocol in the computing platform and informs the MT of the selected paging protocol.
- The network side installs the appropriate interface between the paging protocol and customized paging area selection service.
- The customized paging area selection service component in the MT side retrieves the user profile from Smart Card, and sends this information to the customized paging area selection service component in the network side. The user profile and other location information from the network side are used to decide which algorithm will be used to calculate the customized paging area. E.g. if the MT is installed in the car, the algorithm will be based on the road information and car speed.

The algorithm selected above interacts with some location related applications, e.g. train route service or road map service, to compute customized paging area. The specifications of the customized paging area are sent to both the paging protocol in the network side and the paging protocol in the MT side. This information is used in the network for paging process and is used in MT for location update when the MT crosses the border of the customized paging area.

In addition, if the MT changes its profile, e.g. get off the train, the example will continue from step 1. When the MT leaves the customized paging area, the example will continue from step 3.

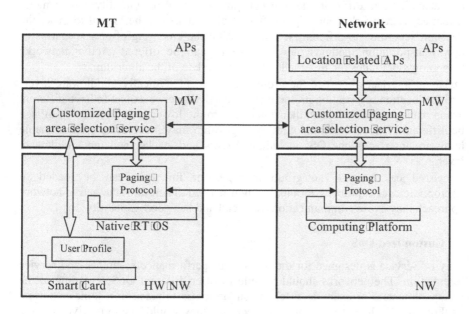

**Fig. 3.** Reference Architecture for Customized Paging Area

Similarly, we can also use sequential paging as the paging protocol. The use of different paging protocols does not only require the capability of loading different components into the computing platform, but also asks for a facility to adapt the interfaces between the MW and the NW layers.

### 4.3 Concatenated Location Management

Currently, location management is designed for individual mobile terminals, provided that each MT moves independently. Such independent location management will need a lot of signaling for handover and location registration. This signaling causes bandwidth waste in the case that MTs are moving in a group, in the same train or car. A novel concept was introduced in [6] to group individual MTs which are moving in the same manner into a given movement class.

The procedure of concatenated location registration is briefly described as follows. Each group of MTs performs a concatenated location registration as a group, and not each individual terminal i.e. only one selected MT in this group needs to register to the network. When a new MT joins the group or when one MT leaves the group, e.g. when entering or leaving a train, the network is informed as well. This registration triggers the network to associate the mobility information of the new MT to the group. There are different ways to organize the group, e.g. by local communication within the group or by some extended node which is in charge of the group management, e.g. which resides in the train. For this procedure, we argue to use active networks technology for the following reasons.

− Flexibility is needed for group management. When and how to form or to release a concatenation is a difficult question but quite essential for mobility management. Using active network technology, different algorithms can be loaded to make the decision according to different scenarios. With the emerging of new scenarios or new transportation methods, new algorithms can be offered. Active networks technology facilitates the flexible loading and update of the algorithm.
− Dynamic interfaces between the group management (typically middleware layer) and the location registration protocol are needed. Since we expect to have different group management solutions, their interfaces to the lower layer protocols can also be different. The implementation of the algorithm may involve the programmable hardware interfaces, the OS, including drivers and stacks, and the middleware layer.
− Localized calculation. The group management functions can be loaded to appropriate access point or terminals when needed. Therefore, the active network approach may avoid communication overhead and has good scalability.

## Customized QoS

Quality of Service is designed for the purpose of performance assurance and service differentiation. The networks should be able to offer different QoS according to the user requirements. For example, a television station is broadcasting some important news. For users with wired networks connections, they would like to receive both the video and audio streams with high quality. For users with their mobile device, only low quality video clips or only audio with text can be received in real time. Also the available wireless capability may change during the connection because of the mobility. E.g. after handover, only lower capacity is available. In this case, the user may prefer some lower quality media service to service interruption. This requires the networks to offer differentiated, dynamically adapted service. In this case, the networks should support multi resolution scalable media.

An active architecture gives a flexible way to scale the media on demand. The MT sends the user scaling preference and radio profile to the access node. Based on this information, the corresponding access node loads suitable media filers or defines the parameters of the filers. A similar idea is implemented in [ ], in which filters are dynamically uploaded to adapt to the capabilities of the visited networking environment.

In mobile and wireless environments, not only the capacity of the connection changes frequently, which make the performance assurance difficult, but also the handoff latency greatly affects the QoS. A lot of approaches are deployed to decrease the handover latency, e.g. fast handover, cellular IP, hierarchical IP, dynamically change the anchor point [ ] [ ]. Different handover strategies have different characteristics. Using active networks, the most suitable handover strategy can be chosen to fulfill the individual QoS requirement of the application.

## Other Application Areas

There exist many other applications of active networks in the area of mobile networks. We list a number of them in the following.

### Hand Over Triggers

An interesting application area is hand over optimization based on layer triggers. In the hand over procedure, a MT first needs to setup a new layer connection, then gets a new address in a foreign network, often called Care of Address, then performs the layer binding update. This procedure causes a hand over latency which might be unacceptable for some delay sensitive applications.

Several handoff schemes have been suggested to eliminate the hand over delay by using cross layer interfaces between layer and layer. One of them is proactive handoff, where link layer triggers assist the MT in determining that a handoff is imminent and establish packet flow to the target AP prior to the handoff event [  ]. This hand over scheme needs the coupling of the layer and layer which fits nicely with our architecture.

### Software Radio

Another important application area is the control of terminal reconfiguration in software radio [ ]. Note that software radio can be applied to both terminals and base stations. Since software radio often aims at simpler forms of software deployment, a local control module is important. This control of the software radio reconfiguration in both base stations and terminals can be provided by active networks. Since new software may implement un anticipated new features, the software reconfiguration process should be as flexible as possible. Hence active networking can be an ideal implementation platform for software radio control mechanisms.

### Ad Hoc Networks

In ad hoc networks, mobile terminals can be networking nodes as well. The connectivity among them may change frequently, depending on their mobility. Therefore, it is difficult to find the best ad hoc routing protocol suitable for all circumstances. For instance, [  ] uses active networks to support customization of routing protocols, where the most suitable routing protocol can be chosen from multiple routing protocols according to the QoS requirements, security concerns, link characteristics, speed of the ad hoc nodes and number of nodes in this ad hoc networks. Since there is potentially a large variety of ad hoc technology and applications, it is important to devise a flexible architecture which is suitable for the integration of ad hoc networks with cellular mobile networks.

## Conclusions

We have discussed active networking and programmability in a complete architecture for future mobile networks. We have shown that a clear common vision is needed to introduce active networking. In particular, we have shown that active networking must consider the configurability of different system components. Another issue is that well accepted standards are needed for a smooth migration to active networks.

We think that future mobile networks can be the ideal target for the adoption of active networks. The main benefit of active networks is the added flexibility. This flexibility is mainly needed for two reasons. First, applications will evolve rapidly and adaptation of lower layer infrastructure will be needed to optimize these applications. Secondly, new wireless technologies and ad hoc networks will require continuous adaptation of the networking layer.

We have shown that flexible cross layer interfaces are needed for these applications. Active networks can be the key enabling technology for this. Furthermore, active networking also supports the separate evolution of different networking layers. This is in contrast to current mobile networks, where often the wireless technology sets the pace for introducing new networking infrastructure.

## References

1. Jamalipour and S. Tekinay, eds., Special issue on Fourth Generation Wireless networks and Interconnecting Standards, October, IEEE Personal Communications Magazine
2. H. Yumiba, K. Imai, and M. Yabusaki, IP Based IMT Network Platform, IEEE Personal Communications Magazine, October
3. N. J. Drew, M. M. Dillinger, Evolution toward reconfigurable user equipment, IEEE Communications Magazine, Vol., Issue, Feb
4. T. Campbell, M. E. Kounavis, and R. R. F. Liao, Programmable Mobile Networks, Computer Networks, Vol., No., pg., April
5. R. Hirschfeld, W. Kellerer, C. Prehofer, H. Berndt, An Integrated System Concept for Fourth Generation Mobile Communication, EURESCOM Summit
6. WWRF, Wireless World Research Forum (WWRF), http://www.wireless-world-research.org
7. M. Christensen, The Innovator's Dilemma, Harward Business Press, Boston
8. M. Bossardt, L. Ruf, R. Stadler, B. Plattner, A Service Deployment Architecture for Heterogeneous Active Network Nodes, Kluwer Academic Publishers, th Conference on Intelligence in Networks, IFIP SmartNet, Saariselkä, Finland, April
9. Brewer, E. A., Katz, R. H., Chawathe, Y., Gribble, S. D., Hodes, T., Giao Nguyen, Stemm, M., Henderson, T., Amir, E., Balakrishnan, H., Fox, A., Padmanabhan, V. N., Seshan, S., A network architecture for heterogeneous mobile computing, IEEE Personal Communications, Volume, Issue, Oct
10. Yue Fang, McDonald, A. B., Cross layer performance effects of path coupling in wireless ad hoc networks, power and throughput implications of IEEE MAC, Performance, Computing and Communications Conference, st IEEE International
11. Raman, B., Bhagwat, P., Seshan, S., Arguments for cross layer optimizations in Bluetooth scatternets, Symposium on Applications and the Internet, Page s
12. Michael Kounavis and Andrew Campbell, Design, Implementation, and Evaluation of Programmable Handoff in Mobile Networks, Mobile Networks and Applications, Kluwer Academic.

[ ] Helen J. Wang, Randy H. Katz, and Jochen Giese, Policy Enabled Handoffs Across Heterogeneous Wireless Networks. In WMCSA. New Orleans, LA.

[ ] Konstantinos P. Psounis, Active Networks: Applications, Security, Safety, and Architectures, IEEE Communication surveys,
http://www.comsoc.org/livepubs/surveys/index.html

[ ] J. Kempf, Dormant Mode Host Alerting ("IP paging") problem statement, RFC, June

[ ] P. Mutaf, C. Castelluccia, DPAC, Dynamic Paging Area Configuration, <draft-mutaf-dpac-.txt> September

[ ] Hsiao Kuang Wu, Ming Hui Jin, Jorng Tzong Horng, Cheng Yi Ke, Personal Paging Area Based On Mobile's Moving Behaviours, IEEE INFOCOM May

[ ] Z. Lei, C. U. Sarazdar, and N. B. Mandayam, Mobility Parameter Estimation for the Optimization of Personal Paging Areas in PCS Cellular Mobile Networks, Proc. 2nd IEEE Signal Processing Workshop on Signal Processing Advances in Wireless Communications (SPAWC' May

[ ] Atsushi Iwasaki, Takashi Koshimizu, Proactive Region Reconfiguration Handover, IPCN April Paris France

[ ] G. Dommety, A. Yegin, C. Perkins, G. Tsirtsis, K. El Malki, M. Khalil, Fast Handovers for Mobile IPv draft-ietf-mobileip-fast-mipv .txt March

[ ] Christian Tschudin, Henrik Lundgren and Henrik Gulbrandsen, Active Routing for Ad Hoc Networks, IEEE Communications Magazine April

[ ] R. Wittmann, M. Zitterbart, *AMnet: Active Multicasting Network*, International Conference on Communications ICC , Atlanta USA June

[ ] D. Decasper, G. Parulkar, S. Choi, J. DeHart, T. Wolf, B. Plattner, *A Scalable High Performance Active Network Node*. In IEEE Network January February

[ ] S. Merugu, S. Bhattacharjee, E. Zegura and K. Calvert, *Bowman: A Node OS for Active Networks*, Proceedings of IEEE Infocom , Tel Aviv Israel March

[ ] David Wetherall, John Guttag and David Tennenhouse, ANTS Network Services without the Red Tape, IEEE Computer April

# Evolution in Action: Using Active Networking to Evolve Network Support for Mobility

Seong-Kyu Song[1], Stephen Shannon[1], Michael Hicks[2], and Scott Nettles[1*]

[1] Electrical and Computer Engineering Department
The University of Texas at Austin
{sksong,shannon,nettles}@ece.utexas.edu
[2] Department of Computer Science
University of Maryland, College Park
mwh@cs.umd.edu

**Abstract.** A key early objective of Active Networking (AN) was to support on-the-fly network evolution. Although AN has been used relatively extensively to build application-customized protocols and even whole networking systems, demonstrations of evolution have been limited.

This paper examines three AN mechanisms and how they enable evolution: active packets and plug-in extensions, well-known to the AN community, and update extensions, which are novel to AN. We devote our presentation to a series of demonstrations of how each type of evolution can be applied to the problem of adding support for mobility to a network. This represents the most large-scale demonstration of AN evolution to date. These demonstrations show what previous AN research has not: that AN technology can, in fact, support very significant changes to the network, even while the network is operational.

## 1 Introduction

The early promise of Active Networking (AN) was two-fold: to introduce computation in the network both for the purpose of application-specific customization and to increase the pace of network service evolution [33, 4]. Thus far there have been numerous demonstrations using AN to customize networks for a specific application's needs (e.g. [21, 10, 22]) as well as some demonstrations of the use of AN to build network infrastructures from scratch [10]. However, while the key mechanisms are in place in many AN systems, there have been few demonstrations of using AN to facilitate network service evolution. One example is the Active Bridge [2], which demonstrated the use of AN to switch between bridging protocols dynamically. Unfortunately, this upgrade was quite simple and did not demonstrate the overall scope of AN possibilities.

---

* This material is based upon work supported by the National Science Foundation under Grant No. CSE-0081360. Any opinions, findings, and conclusions or recommendations expressed in this material are those of the author(s) and do not necessarily reflect the views of the National Science Foundation.

J. Sterbenz et al. (Eds.): IWAN 2002, LNCS 2546, pp. 146–161, 2002.

We hope to fill this gap in the research record by showing how AN can be used to upgrade a network's services on the fly, without centralized coordination and without halting network service. By doing so, we are making a strong claim that AN *can* be used to quicken the pace of network service evolution. For our demonstration, we present two main examples. First, we show how to augment an active network that provides standard, IP-like service to support routing for mobile hosts, in the spirit of Mobile-IP [29, 30]. Second, we show how to augment an active, mobile ad-hoc network with support for multi-hop routing. While our primary goal is to demonstrate the capabilities of AN technology in evolving a network, a secondary goal is to explore the suitability and usefulness of AN techniques within the mobile networking domain.

These demonstrations have had the side-benefit of helping us better understand the types of evolution that AN's can and should support, and the mechanisms required to implement them. In particular, we have identified three programmability mechanisms—active packets (APs), plug-in extensions, and update extensions—which in turn enable three different types of evolution. APs and plug-in extensions have been well studied in AN research (e.g. [32, 36, 8, 28, 35, 14, 24, 3, 37, 20, 5, 6]) but update extensions have not been explored for AN [12].

A key differentiating factor among these mechanisms concerns whether a new service is application-aware or application-transparent. Application-aware network services require that for an application to use a new service, it must be aware that it is doing so. For example, using IP-style multicast requires the sending application (or perhaps the middle-ware used by the application) to send to a special multicast address. In contrast, application-transparent services are those that act without the application's knowledge. For example, in IP-style mobility, packets destined for a host's home network are transparently forwarded to that host's current remote network; the sending application does not need to be aware of mobile IP services for them to work. In making this distinction, we have realized that APs and, in many cases, plug-in extensions cannot solely provide transparent service; they require the aid of update extensions. However, the added power of plug-in and update extensions makes them a greater security risk, implying that services would benefit from using a combination of mechanisms to balance the needs of the application and of the network.

We hope that this paper will raise the level of discussion on the mechanisms required by active networks to truly support not just application-level customization, but true network service evolution. We begin in Section 2 by describing the three types of network evolution and their enabling AN technologies. In Section 3 we discuss our implementation testbed, MANE, and its programmability mechanisms. We also provide enough background on mobile-IP-style mobility and ad-hoc networks for the reader to understand our demonstrations. In Sections 4, 5 and 6 we subsequently present our demonstrations of each of the three types of network evolutions described in Section 2. Finally, in Section 7, we summarize some of the lessons learned from our case studies and conclude.

## 2    Network Evolution

Broadly speaking, by "network evolution" we mean any incremental change to a network that modifies or enhances existing functionality or adds new functionality. In the context of Active Networking a somewhat more ambitious goal is appropriate: evolution should be able to occur at remote nodes while the network is operational with only minimal disruption to existing services.

AN achieves evolution by changing the programs that operate the network. Thus the ways in which we can evolve the network are dictated by the programmability mechanisms that are available to make such changes. In some cases, these mechanisms are AN specific, but generally they are drawn from general programming language technology. Thus, although later we will choose instances of these mechanisms that are specific to our platform, this discussion is general and applies broadly to AN systems.

In this section, we describe three mechanisms for achieving AN evolution. In each case, we discuss what type of evolution is supported by the mechanism. We also consider how the mechanism might support application aware or transparent evolutions. Later in the paper we will show concrete examples of evolutions of each type using the mechanisms outlined here.

### 2.1    Active Packets

Active packets (AP) are perhaps the most radical AN technology for evolution and they are the only mechanism that, at a high-level, are specific to AN. Such packets carry (or literally are) programs that execute as they pass through the nodes of network. A packet can perform management actions on the nodes, effect its own routing, or form the basis of larger protocols between distributed nodes, e.g. routing protocols. Such packets can form the glue of the network, much like conventional packets, but with qualitatively more power and flexibility.

The AN community has explored a number of AP systems. The early systems include Smart Packets [32], ANTS [36], and PLAN [8], while more recent systems include PAN [28], SafetyNet [35], StreamCode [14], and SNAP [24]. Although these systems differ on many details of their design and implementation, they all support the basic AP model and thus the same general styles of evolution.

Active packets support the first and simplest type of network evolution we identify, *Active Packet evolution*, which does not require changes to the nodes of the network. Instead, it functions solely by the execution of APs utilizing standard services. The disadvantage of this approach is that taking advantage of new functionality requires the use of new packet programs. This means that at some level the applications using the functionality must be aware that the new functionality exists. This is the kind of evolution facilitated by pure AP systems, such as ANTS [36] and in essence it embodies the AN goal of application-level customization. We will demonstrate how this kind of evolution can be used to implement application-aware mobile-IP style mobility in Section 4.

## 2.2    Plug-In Extensions

The programmability mechanism that is broadly familiar outside the AN community is the *plug-in extension*. Plug-in extensions can be downloaded and dynamically linked into a node to add new node-level functionality. For this new functionality to be used, it must be callable from some prebuilt and known interface. For example, a packet program will have a standard way of calling node resident services. If it is possible to add a plug-in extension to the set of callable services (typically by extending the service name space) then such an extension "plugs in" to the service call interface.

Plug-in extensions are commonly used outside of AN. For example, the Linux kernel enables plug-in extensions for network-level protocol handlers, drivers, and more. Java-enabled web browsers support applets, which are a form of plug-in extension. Plug-ins are also common to AN. In CANES [3], nodes execute programs that consist of a fixed underlying program and a variable part, called the injected program. The fixed program contains *slots* that are filled in plug-in extensions. In Netscript [37], programming takes place by composing components into a custom dataflow. In this case, each element in the composed program is a plug-in, and the abstract description of such an element forms the extension interface. Plug-ins are used in hardware-based approaches as well, including the VERA extensible router at Princeton [20], and Active Network Nodes (ANN) at Washington University and ETH [5, 6].

Plug-in extensions support the second type of evolution we identify, *plug-in extension evolution*. When used in conjunction with APs, packet programs can use new node resident services specialized to their needs rather than just standard services. Such evolution is particularly important if standard services are not sufficient to express a needed application. The combination of Active Packet and plug-in extension evolution is facilitated by systems such as PLANet [10], ALIEN [1], and SENCOMM [17]. We will demonstrate how this kind of evolution can be used to add multi-hop routing to an ad-hoc mobile network in Section 5.

Plug-in extensions that must be referenced by new AP programs are obviously not application transparent. However, as long as a plug-in simply replaces an existing interface, whether that interface is accessed from an AP or even in a more conventional system that does not support APs at all, then the evolution can be application transparent. This situation in occurs in CANES, for example. However, the system still must have been designed to allow the required change (e.g. in CANES, this is made possible by the slots in the fixed program).

## 2.3    Update Extensions

The final programmability mechanism we consider is the *update extension*. Update extensions may also be downloaded, but they go beyond plug-in extensions in that they can update or modify existing code and can do so even while the node remains operational. Thus, such extension can add to or modify a system's functionality even when there does not exist an interface for it to hook into.

There is significant research literature on such extensions (e.g. [7, 23, 15] to name a few) although in general the focus has been on code maintenance rather than evolving distributed system functionality. The specific system we are using [11] was initially inspired by the difficulties of crafting a plug-in interface for the packet scheduler in PLANet [10, 12]. However, the system itself is not specific or specialized to AN.

Update extensions support the final type of network evolution we identify, *update extension evolution*, which occurs when network nodes are updated in more or less arbitrary ways. This means that the evolution can affect the operation of existing functionality, even if such functionality was not explicitly designed to be extended (as was required for plug-in extensions). This means that in general evolutions that are transparent to the clients of a service are feasible. To our knowledge, only our current system, MANE, supports this type of network evolution. We will demonstrate how this kind of evolution can be used to implement application-transparent, mobile-IP-style mobility in Section 6.

# 3    Technology

Before presenting our examples, we discuss the key technologies used in their implementation. We discuss our Mobile Active Network Environment (MANE), and the details of how it provides the key evolutionary mechanisms, as well as the networking protocols we will evolve MANE to support.

## 3.1    MANE

MANE is the logical descendant of our previous AN testbed, PLANet [10], with improvements in a number of areas. MANE is very loosely based on the limited initial prototype described by Hornof [16].

MANE implements all three evolutionary mechanisms. AP programs are written in a special-purpose language, PLAN [8]. MANE nodes and their extensions (both plug-in and update) are implemented in software based on Typed Assembly Language (TAL) [26]. TAL is a cousin of proof-carrying code (PCC) [27], a framework in which native machine code is coupled with annotations such that the code can be automatically proved to satisfy certain safety conditions. A well-formed TAL program is memory safe (i.e. no pointer forging), control-flow safe (i.e. no jumping to arbitrary memory locations), and stack-safe (i.e. no modifying of non-local stack frames) among other desirable properties. TAL has been implemented for the Intel IA32 instruction set; this implementation, called TALx86 [25], includes a TAL verifier and a prototype compiler from a type-safe C-like language called Popcorn, to TAL. MANE is actually written in Popcorn.

Enhancing a node with a plug-in or update extension is achieved through type-safe dynamic linking [13]. A frequent use of plug-in extensions is to extend the services available to PLAN packets. To do this, extensions are loaded and plugged into the service symbol table. When future APs are processed, they will reference this table, and thus have access to the new functionality. Update extensions are dynamically linked as well, but differ from plug-in extensions in that the existing node code and any existing extensions are *relinked* following the update [11]. In this way, existing code 'notices' that a new version of a particular module has been loaded the next time that code is called (and the old code will run until this point). Care must be taken when using this technology to avoid errors arising from the timing of an update [11].

MANE presents a two-level namespace architecture. References in the packet to services are resolved by the service plug-in namespace, while references between plug-ins and/or the rest of the program are resolved by the program namespace. In both cases, these namespaces may be changed at runtime to refer to new entities. A benefit of this separation is that the presentation of each namespace can be parameterized by policy, for example to include security criteria. This is useful because the division between the Active Packet, plug-in extension, and the update extension layers constitutes a likely division of privilege. APs are quite limited in what they can do, so we allow arbitrary users to execute those packets. However, when a packet calls a service, implemented as a plug-in extension, the privilege of the packet can be checked before allowing the call to take place [9]. Similarly, when an update extension is loaded, the privilege of those extensions that would relink against the update extension can be checked before allowing the relinking to take place.

The basic network model of MANE is much like an IP network. MANE addresses are globally unique and hierarchical. The hierarchy is based on sub-nets of nodes and individual nodes on a sub-net can broadcast to each other, while communication with nodes on other networks must be mediated by routers. MANE routers run a conventional link-state routing protocol. MANE supports a form of DHCP, which can dynamically assign both an address and a default router to a node connected to a given network. MANE uses an ARP-style protocol to resolve the link-level address corresponding to a network level address and there is a provision for proxy-ARP as well.

There are also key differences between an IP network and MANE. MANE communicates using only APs and nodes can be extended and updated. To support APs, MANE provides certain basic services, such as means to identify a node and to store and retrieve *soft state* based on a key. A soft-store is an essential service for APs and is provided by many AN systems (e.g. [36, 22, 24, 10, 28]).

At its lowest level, MANE emulates broadcast networks by keeping track of which nodes are on a particular sub-net and using UDP to communicate between neighbors. Broadcast is achieved by repeated unicast. This level also supports emulation of physical node mobility, allowing a node to leave a sub-net and to join new sub-nets. To the higher-level software, this emulation is transparent.

## 3.2    Mobility and Mobility Protocols

Several reasons motivated our choice of mobility from which to draw our examples. First, mobility is an area in which new protocols and improvements to existing protocols are being developed rapidly. Thus it is an area where better evolutionary capabilities could be a real benefit, since then protocols could be deployed and later upgraded and replaced as new techniques develop. Second, in the particular area of mobile-IP, current protocols are constrained in their design to require only local changes to the network infrastructure. Practical evolution capability would allow other (preferable) protocols to be developed. Finally, mobility is an interesting domain in its own right, and the current work allows us to begin to understand the issues there in the context of our design and implementation environment.

The goal of the work here is not to innovate in the area of mobility. By choice, we have taken existing, well understood mobile protocols and shown how networks can be evolved to support them. We have done this both to gain an understanding of these protocols, but more importantly to avoid clouding the evolution issues with issues involving new mobile protocols. We make no claims that the protocols we use here are the best possible ones, but rather they are known solutions that serve to illustrate how wide classes of protocols can be easily evolved and integrated using AN techniques. For broader considerations about how AN could be applied to mobile networks, see [34, 31].

**Mobile-IP** To demonstrate both Active Packet and update extension evolution, we have chosen to add support to MANE for what is essentially mobile-IP [29, 30]. Only end nodes can be mobile. A mobile host has a "home" network, which is implicit in its address. Even when a host is mobile, packets for it are sent to its home network for delivery. If a host is not mobile, packets are delivered in the normal way. If a host is mobile, when it connects to a remote or "foreign" network, it uses DHCP to acquire a local address, allowing the node to function as its own foreign agent. The mobile host then sends a registration packet to one of the routers on its home network with the information that it can be contacted at its newly acquired address. This router acts as the "home agent". When a packet arrives at the home agent destined for the mobile host, it is tunneled to the mobile host using its address on the foreign network. There the packet can be removed from the tunnel and delivered.

Consider what must be added to MANE to support this protocol:

1. The home agent must be identified.
2. There must be a way to send a registration packet.
3. There must be a way to recognize when a packet arrives at the home agent.
4. There must be a way to create a tunnel.
5. There must be a way to remove the original packet from the tunnel.

The application-aware and transparent evolutions will share the same implementation for many of these functions. The shared implementation is the inherently non-transparent part of mobile-IP, including basically all but point 3.

**Multi-hop Ad-Hoc Routing** To demonstrate plug-in extension evolution, we have chosen to show how a multi-hop routing protocol can be added to an ad-hoc mobile network with only neighbor-to-neighbor communication. The protocol we have chosen is Dynamic Source Routing (DSR) [19, 18]. While a variety of protocols could also be deployed using our techniques, DSR has the advantage of being simple and well understood.

DSR is an on-demand routing protocol, which searches for a source route to some destination by flooding a ROUTE REQUEST packet on the network when the initiating node has data packets to send. This ROUTE REQUEST packet is forwarded on the whole network to the target, which replies with a ROUTE REPLY packet. When the ROUTE REPLY packet arrives, the initiator can start sending data packets using the source route found in the ROUTE REPLY packet. It also saves the route in its route cache. An important optimization is for intermediate nodes to also maintain a route cache so that if they already know a route to the requested destination, they can simply send a ROUTE REPLY, thus limiting flooding. Data in the route caches times out, to avoid stale routes.

Consider what needs to be added to MANE to support this protocol:

1. ROUTE REQUEST packets must be flooded. This includes the need to check if an intermediate node has already seen the particular flood message.
2. There needs to be a route cache that can be queried and updated.
3. There needs to be a ROUTE REPLY packet that carries the source route back to the initiator.

Note that many of the preexisting capabilities of MANE are not needed by DSR, mostly just the ability to send packets one hop and to name nodes.

## 4   Active Packet Evolution

If it is acceptable for the node trying to communicate with the mobile host to be aware that the host might be mobile, then it is possible to implement the basic mobile-IP protocol discussed in Section 3 using only APs. Some of this implementation can be shared with the update extension evolution example (in Section 6), we discuss this first, followed by a discussion of the aspects unique to APs.

### 4.1   Setting the Forwarding Path

The application-aware and transparent versions share the same infrastructure for setting up the forwarding path to a mobile host (while they differ in how packets are actually forwarded). Before a mobile node leaves its home network, it must identify the router that serves as its home agent. For simplicity, we assume that its default router serves this purpose. Once the node has attached itself to a new network and has a unique address, it sends an AP containing a control program to register itself to its home agent. When executed, this program simply adds the information to the home agents soft-state keyed by the mobile nodes home network address. Both the application aware and transparent versions share the same soft-state entries, allowing them to use the same control program and to co-exist.

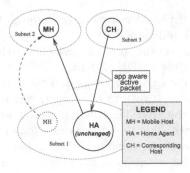

**Fig. 1.** Active Packets for Mobile-IP

## 4.2   Forwarding: The Application-aware Protocol

The key questions remaining are how do we detect that a packet is at the home agent of a mobile host and how is the packet then tunneled to the unique address. Because this version is application aware, both of these steps can be done by having the application use a special AP as shown in Figure 1.

The PLAN code for the packet that must be sent by the application is shown in Figure 2. GetToAgent is the main function and when it executes, it first looks up dest in the soft-store using lookupTuple. If that succeeds, it has found the home agent and it uses OnRemote to send a new packet, the tunnel, that will execute FoundFA at the foreign agent with the same arguments as getToAgent. OnRemote provides multi-hop transmission of the packet with out execution until it reaches the foreign agent. If the lookup fails the handle will execute. If we have actually reached the host then we deliver the packet. Otherwise, it looks up the next hop toward dest. It then uses OnNeighbor, which only transmits a packet one hop, to send the packet. Thus the packet travels hop-by-hop looking for the home agent.

Now consider the FoundFA function. It executes on the foreign agent, which is in our case is the mobile host, but might be some other node on the same sub-net. It sends a packet to the dest that does the deliver. This is where the original packet is removed from the tunnel. Notice that all of that functionality is encoded in the tunnel packet program itself, the foreign agent does not need to have any knowledge of its role as a tunnel endpoint, it just has to support PLAN.

## 5   Plug-In Extension Evolution

Some evolutions do not just augment or customize some existing aspect of the network, but rather add wholly new functionality. Such evolutions may be problematic with just APs because of their 'fixed vocabulary.' To rectify this problem, we can load new services in the form of plug-in extensions to extend a packet's

```
fun getToAgent(dest, payload, port) =
 try
  let val fagent = lookupTuple(dest) in
   OnRemote(|FoundFA|(dest, payload, port), fagent, getRB(), defRoute)
  end
 handle NotFound =>
  if (thisHostIs(dest)) then deliver(payload, port)
  else
   let val next = defaultRoute(dest) in
    OnNeighbor(|getToAgent|(dest, payload, port), #1 next, getRB(),
    #2 next))
   end

fun FoundFA(dest, payload, port) =
 let val hop = defaultRoute(dest) in
  OnNeighbor(|deliver|(payload, port), #1 hop, getRB(), #2 hop))
 end
```

**Fig. 2.** PLAN packet for Mobile-IP

**Algorithm 5.1:** ROUTE DISCOVER(*Simple DSR*)

**procedure** ROUTE REQUEST(*Target, RouteRec*)
  **if** *Duplicate Request Packet  Or  My Address Already In Route Rec*
  **then** *Discard*
  **else** $\begin{cases} \textbf{if } \textit{This Host Is Target} \\ \quad \textbf{then } \textsc{Route Reply}(\textit{RouteRec}) \\ \quad \textbf{else } \begin{cases} \textit{Append My Address To Route} \\ \textit{Flood To All Neighbors} \end{cases} \end{cases}$

vocabulary, and thus enable implementation of the new networking functionality. More generally, if a node defines an extensible interface, new extensions can be loaded and plugged into this interface to provide extended or enhanced functionality. This is the essence of plug-in extension evolution.

As we will show, it is possible to add multi-hop routing based on DSR to an ad-hoc network only supporting neighbor communication by using plug-in extension evolution. The main functionality is embodied in a special AP for communicating routes between nodes, which makes use of a non-standard service. This packet does not need to be sent by an application, but rather by the networking software on the ad-hoc node, and therefore the dynamic deployment of the protocol does not require applications to be aware of the change.

## 5.1   An Active Packet for Route Discovery

Consider the pseudocode shown in Algorithm 5.1 for a simple AP that implements route discovery. The packet itself embodies many key aspects of the protocol directly. In particular, it does duplicate elimination, tests for routing loops, detects termination and sends a reply, and performs flooding. In general, since many protocols have relatively simple control flow, simple APs can implement key aspects of the protocol directly. However, this algorithm benefits from node-resident services that are specific to the protocol, particularly to detect if this is a duplicate request or if the current node is already in the route record.

## 5.2  Deploying DSR Dynamically

Implementing the required node-resident services is straightforward. To deploy the service dynamically, we need a way to discover the nodes that will need the service, upload the extension to those nodes, and install it. Since a general treatment of node-resident service discovery and deployment is a topic of research, consider the following basic idea. One way to discover which nodes will need an extension is to require that when a packet arrives at a node, it ensures that extensions it needs are installed before it executes. Acquiring a needed extension could be done in a number of ways. For example, it could always carry the extension with it, it could retrieve the extension from a well-known code repository, or it could use an ANTS-style distribution protocol [36] in which the extension is obtained from the source of the current packet. For our demonstration we chose the simplest option—carrying the extension in the route request packet— but any of these solutions could be implemented in MANE. The key point is that given the ability to dynamically load code, the network itself provides the mechanisms to transport loadable code to the locations where it is needed. The pseudocode for the simple approach is shown in Algorithm 5.2.

**Algorithm 5.2:** ROUTE DISCOVER($SimpleDSR + DynamicLoading$)

> **procedure** ROUTE REQUEST($Target, RouteRec, Extension$)
>   **if** $DSR\ Service\ Not\ Present$
>     **then** $Load\ DSR\ Extension\ From\ This\ Packet$
>   $Same\ Code\ as\ Above$

# 6  Update Extension Evolution

A potential problem with the evolutions shown so far is that some part of the system must send special packets to take advantage of the new service. In some applications, being aware of new packets or services is acceptable, while in others it is not. The latter is true for mobile IP: it would be unreasonable to change all possible senders on the network to use our special packets from Section 4 to send to a potentially mobile host. In this section, we demonstrate how using update extensions, we are able to evolve the network so that forwarding is transparent to the sender and does not require using a special packet.

The basic strategy is shown Figure 3. Here a packet that is not aware of mobility is destined for a mobile node. However, because we have updated the Home Agent to support transparent forwarding, it is able to intercept the packet and tunnel it to the mobile host. Thus, although we use Active Packets and

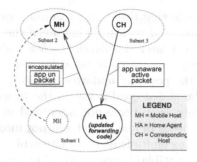

**Fig. 3.** Update extension Mobile-Ip evolution

plug-in extensions to help perform our evolution in a convenient way, the key to transparent evolution is really the use of update extensions.

Because mobility is inherently not transparent to the mobile host itself, we can reasonably have it set up the forwarding path to the remote agent as described in Section 4.1, with the added benefit that the nontransparent and transparent techniques can coexist.

## 6.1   Forwarding: The Application-Transparent Protocol

To make the forwarding transparent to the sender requires a way to detect that a packet has arrived at the home agent and to forward it to the foreign agent *without having to rewrite the sender's packet code*. The most straightforward way of doing this is to modify the router's forwarding logic: whenever a packet arrives, look up its destination address in the soft state that is used to record which hosts are mobile, and if present, forward the packet to the foreign agent. Pseudocode for router forwarding logic is shown below (in a C-like syntax), with the additional part shown in italics:

```
void sendToNextHop(pkt_t packet, host_t dest){
  host_t nextHop = Route(dest);
  if is_mobile_host(dest) then
    tunnel_to_foreign_agent(packet, dest);
  else
    send_on_link(packet, dest, next_hop);
}
```

Note that the code implementing is_mobile_host and tunnel_to_foreign_agent would be implemented elsewhere and loaded separately.

While the addition of an if-statement to the forwarding loop is conceptually simple, it is impossible to realize *on-the-fly* without the support of update extensions. This is because the forwarding loop in MANE was not designed for change; that is, it did not provide a plug-in interface for performing new operations in the loop.[1] As a result, effecting this change would require changing the code statically and recompiling, and then bringing down the node and restarting it with the new code. This is practical when only a few nodes need to be updated, but much less so if an evolution needs to be widespread.

On the other hand, the power of update extensions makes them more dangerous. For example, the added conditional in the forwarding loop above will be invoked for *all* packets, even those not interested in mobility. In an active packet-only system, the needs of one packet will not interfere with another's in this way. Similarly, allowing multiple users update arbitrary parts of the router's code could result in incompatible changes, and/or an unintelligible code base. As such, update extensions will likely be limited to privileged users, limiting their applicability.

Implementing this change as an update extension in MANE requires two actions. First, dynamically load and link a mobility module that implements the test of whether a packet needs to be forwarded as well as providing the tunneling code. Second, dynamically load a new version of the forwarding code that does the required test and then update the old running code with the new version.

One interesting point remains. How is the tunnel itself created? In essentially the same way as in the non-transparent case, although obviously done by the node resident mobility code. A new AP is created which when executed on the foreign agent unpacks the original packet and delivers it. Note that tunneling this way works even when the packet being tunneled is not active and thus again avoids the need for the foreign agent to act explicitly as the tunnel end-point.

# 7    Conclusions

To conclude, we summarize the lessons learned while demonstrating each of the three kinds of evolution.

*Active Packet Evolution* The chief advantage of Active Packet evolution is that it is lightweight and allows third parties to enhance the functionality of the network without changing the nodes themselves. Thus, from the point of view of security, this style of evolution is the most desirable and gives the widest variety of users the ability to evolve the network. One disadvantage is that it is inherently not application transparent. Another key disadvantage is that if the existing node interface does not support some critical piece of functionality, it may be impossible to achieve the desired result. Despite this, our example (and others, e.g. [22, 21]) shows that even with only very basic services, non-trivial applications are feasible. An interesting challenge to the AN community is to design a set of node-resident services that maximizes the range of evolutions that can be achieved with just APs. Since, as in our example, the APs can

---

[1] We could imagine designing the forwarding loop to allow for extensibility, as is the case in CANES [3]. However, there will always be parts of the system that were not coded to anticipate future change, and therefore will lack a plug-in interface. As a result, these parts of the system can only be updated if with update extensions.

often embody a substantial part of the control aspect of a protocol, this effort would be quite different from typical protocol design and would need to focus on providing the generic components that support the aspects of a variety of protocols that can not be expressed in the packets. Interestingly, just the simple soft state provided by ANTS [36] and our system, is already a significant step in that direction.

*Plug-in Evolution* Plug-in extensions provide significantly greater evolution possibilities when used with APs. This is because protocols can have both new packet programs and new node resident components. This power comes directly from having the service namespace in the AP system provide a generic plug-in interface. One obvious disadvantage is that such service plug-ins are not application transparent.

Our examples did not take advantage of plug-in extensions that were not accessed from packet programs, as is possible in CANES [3], and Netscript [37], among others. Such plug-ins have the advantage that they can be used to achieve application-transparent evolution, but only in ways that the underlaying system has anticipated by providing a plug-in interface.

A general disadvantage of plug-in evolution is that dynamically loading plug-ins implemented using general purpose programming language code poses a significant security risk. This will restrict the range of users who can deploy these kinds of extensions. This suggests another research opportunity for the AN community, designing plug-in extensions that can be safely loaded by third parties. One approach to this would be to use special-purpose languages with built-in security properties, in the spirit of PLAN.

*Update Evolution* The example here is the first example of update evolution in AN, chiefly because MANE is the first AN system to support such evolution. The advantage of this approach is clear: application-transparent evolution can be achieved even when the system design has not anticipated the need for a particular kind of change. In some sense, this embodies the entire goal of AN. There are two disadvantages. One, dynamic updating is not a widespread technology like dynamic loading, though it can be conceptually simple to implement [11]. Second, and more importantly, the power of update extensions implies the need for greater security and reliability considerations than for plug-in extensions or active packets. In the short term, this means that only privileged users with access to the entire router code base should make use of this technology. In the long term, more research is needed to understand how to manage multiple updaters of the same code, and ways to limit their system-wide effects.

We have presented a taxonomy of types of AN evolution driven by the underlaying language technology. Based on that taxonomy, we presented a series of examples that illustrate what expressibility gains are possible as successively more powerful techniques are used. However, these gains in expressibility are balanced by the greater security risks of more powerful techniques. Greater security risks imply that fewer users may deploy a system. Thus a basic design principle for AN systems should be to use the least powerful evolutionary mechanisms possible so as to maximize the range of users that may deploy a system.

# References

[1] D. S. Alexander, W. A. Arbaugh, A. D. Keromytis, and J. M. Smith. A Secure Active Network Environment Architecture: Realization in SwitchWare. *IEEE Network Magazine*, 12(3):37–45, 1998. Special issue on Active and Controllable Networks.

[2] D. S. Alexander, M. Shaw, S. M. Nettles, and J. M. Smith. Active Bridging. In *Proceedings, 1997 SIGCOMM Conference*. ACM, 1997.

[3] S. Bhattacharjee. *Active Networking: Architecture, Compositions, and Applications*. PhD thesis, Georgia Institute of Technology, August 1999.

[4] K. Calvert, S. Bhatacharjee, E. Zegura, and J. P. Sterbenz. Directions in Active Networks, October 1998.

[5] D. Decasper, Z. Dittia, G. M. Parulkar, and B. Plattner. Router plugins: A software architecture for next generation routers. In *SIGCOMM*, pages 229–240, 1998.

[6] D. Decasper, G. Parulkar, S. Choi, J. DeHart, T. Wolf, and B. Plattner. A scalable, high performance active network node, 1999.

[7] O. Frieder and M. E. Segal. On Dynamically Updating a Computer Program: From Concept to Prototype. *Journal of Systems and Software*, 14(2):111–128, September 1991.

[8] M. Hicks, P. Kakkar, J. T. Moore, C. A. Gunter, and S. Nettles. PLAN: A Packet Language for Active Networks. In *Proceedings of the Third ACM SIGPLAN International Conference on Functional Programming*, pages 86–93. ACM, 1998.

[9] M. Hicks, A. D. Keromytis, and J. M. Smith. A Secure PLAN (Extended Version). In *Proceedings of the DARPA Active Networks Conference and Exposition (DANCE)*. IEEE, May 2002.

[10] M. Hicks, J. T. Moore, D. S. Alexander, C. A. Gunter, and S. Nettles. PLANet: An Active Internetwork. In *Proceedings of the Eighteenth IEEE Computer and Communication Society INFOCOM Conference*, March 1999.

[11] M. Hicks, J. T. Moore, and S. Nettles. Dynamic Software Updating. In *Proceedings of the ACM SIGPLAN Conference on Programming Language Design and Implementation*, pages 13–23. ACM, June 2001.

[12] M. Hicks and S. Nettles. Active Networking means Evolution (or Enhanced Extensibility Required). In *Proceedings of the Second International Working Conference on Active Networks*, October 2000.

[13] M. Hicks, S. Weirich, and K. Crary. Safe and Flexible Dynamic Linking of Native Code. In *Preliminary Proceedings of the ACM SIGPLAN Workshop on Types in Compilation*, Technical Report CMU-CS-00-161. Carnegie Mellon University, September 2000.

[14] K. Hino, T. Egawa, and Y. Kiriha. Open programmable layer-3 networking. In *Proceedings of the Sixth IFIP Conference on Intelligence in Networks (SmartNet 2000)*, September 2000.

[15] G. Hjálmtýsson and R. Gray. Dynamic C++ Classes, A lightweight mechanism to update code in a running program. In *Proceedings of the USENIX Annual Technical Conference*, June 1998.

[16] L. Hornof. Self-Specializing Mobile Code for Adaptive Network Services. In *Proceedings of the Second International Working Conference on Active Networks*, volume 1942 of *Lecture Notes in Computer Science*. Springer, 2000.

[17] A. W. Jackson, J. P. Sterbenz, M. N. Condell, and R. R. Hain. Active Monitoring and Control: The SENCOMM Architecture and Implementation. In *Proceedings of the DARPA Active Networks Conference and Exposition (DANCE)*, May 2002.

[18] D. Johnson, D. Maltz, and J. Broch. DSR: The Dynamic Source Routing Protocol for Multihop Wireless Ad Hoc Networks. In C. E. Perkins, editor, *Ad Hoc Networking*. Addison-Wesley, 2001.

[19] D. B. Johnson and D. A. Maltz. Dynamic source routing in ad hoc wireless networks. In Imielinski and Korth, editors, *Mobile Computing*, volume 353. Kluwer Academic Publishers, 1996.

[20] S. Karlin and L. Peterson. VERA: an extensible router architecture. *Computer Networks (Amsterdam, Netherlands: 1999)*, 38(3):277–293, 2002.

[21] U. Legedza, D. Wetherall, and J. Guttag. Improving the Performance of Distributed Applications Using Active Networks. In *IEEE INFOCOM*, March 1998.

[22] L. Lehman, S. Garland, and D. Tennenhouse. Active Reliable Multicast. In *IEEE INFOCOM*, March 1998.

[23] S. Malabarba, R. Pandey, J. Gragg, E. Barr, and J. F. Barnes. Runtime support for type-safe dynamic Java classes. In *Proceedings of the Fourteenth European Conference on Object-Oriented Programming*, June 2000.

[24] J. T. Moore, M. Hicks, and S. Nettles. Practical programmable packets. In *Proceedings of the 20th Annual Joint Conference of the IEEE Computer and Communications Societies*, April 2001.

[25] G. Morrisett, K. Crary, N. Glew, D. Grossman, R. Samuels, F. Smith, D. Walker, S. Weirich, and S. Zdancewic. TALx86: A Realistic Typed Assembly Language. In *Second Workshop on Compiler Support for System Software*, Atlanta, May 1999.

[26] G. Morrisett, D. Walker, K. Crary, and N. Glew. From System F to Typed Assembly Language. *ACM Transactions on Programming Languages and Systems*, 21(3):527–568, May 1999.

[27] G. Necula. Proof-Carrying Code. In *Twenty-Fourth ACM Symposium on Principles of Programming Languages*, pages 106–119, Paris, Jan. 1997.

[28] E. Nygren, S. Garland, and M. F. Kaashoek. PAN: A high-performance active network node supporting multiple mobile code systems. In *OPENARCH'99*, March 1999.

[29] C. Perkins. IP mobility support. Internet RFC 2002, October 1996.

[30] C. Perkins. IP mobility Support Version 2. Internet Draft, Internet Engineering Task Force, Work in progress., 1997.

[31] B. Plattner and J. P. Sterbenz. Mobile wireless activenetworking: Issues and research agenda. In *IEICE Workshop on Active Network Technology and Applications (ANTA) 2002*, Tokyo, March 2002.

[32] B. Schwartz, A. W. Jackson, W. T. Strayer, W. Zhou, R. D. Rockwell, and C. Partridge. Smart packets: Applying active networks to network management. *ACM Transactions on Computer Systems*, 18(1), February 2000.

[33] D. L. Tennenhouse, J. M. Smith, W. D. Sincoskie, D. J. Wetherall, and G. J. Minden. A Survey of Active Network Research. *IEEE Communications Magazine*, 35(1):80–86, January 1997.

[34] C. Tschudin, H. Lundgren, and H. Gulbrandsen. Active Routing for Ad Hoc Networks, April 2000.

[35] I. Wakeman, A. Jeffrey, T. Owen, and D. Pepper. SafetyNet: A language-based approach to programmable networks. In *OPENARCH'00*, April 2000.

[36] D. J. Wetherall, J. Guttag, and D. L. Tennenhouse. ANTS: A Toolkit for Building and Dynamically Deploying Network Protocols. In *IEEE OPENARCH*, April 1998.

[37] Y. Yemini and S. daSilva. Towards programmable networks, 1996.

# AMnet 2.0: An Improved Architecture for Programmable Networks

Thomas Fuhrmann, Till Harbaum, Marcus Schöller, and Martina Zitterbart

Institut für Telematik
Universität Karlsruhe, Germany

**Abstract** AMnet 2.0 is an improved architecture for programmable networks that is based on the experiences from the previous implementation of AMnet. This paper gives an overview of the AMnet architecture and Linux-based implementation of this software router. It also discusses the differences to the previous version of AMnet. AMnet 2.0 complements application services with net-centric services in an integrated system that provides the fundamental building blocks both for an active node itself and the operation of a larger set of nodes, including code deployment decisions, service relocation, resource management.

**Keywords:** Programmable Networks, Active Nodes

## 1 Introduction

The idea of active and programmable networks has been studied for several years already. But despite its great flexibility and potential, practical restrictions so far kept this idea from actual deployment. We believe that especially the tradeoff between general programmability and security considerations proves to be a major obstacle for actual deployment. Generally programmable active nodes have to offer full access to the network traffic, including inspection and modification of all the data routed through that node. Thus, deficient or malicious code loaded into such a node can threaten the network's integrity. Closely related to these security considerations is the lack of a sufficient amount of usage scenarios in which active and programmable networks provide an advantage over less precarious approaches like proxies and application-layer overlay-networks. Since we believe that both aspects have to be addressed in their mutual relation, the redesign of the AMnet architecture was guided by the consideration of usage scenarios for active and programmable networks that go beyond the mere support of heterogeneous group communication in AMnet 1.0.

The previous concept of AMnet [14, 7, 8] mainly pursued the scenario of a sender based heterogeneous multicast where different end-systems have different requirements. Following the multicast design principle, adaptation of the data was pushed from the sender to some node along the distribution tree. Signaling relied on session and service announcements conveyed along the distribution tree. Set-up of a service was solely triggered by the receivers. Thus, the application of AMnet 1.0 required both, sender and receiver, to be adapted to AMnet.

J. Sterbenz et al. (Eds.): IWAN 2002, LNCS 2546, pp. 162–176, 2002.

The new AMnet 2.0 concept, in contrast, extends the original scenario to cases where the sender, the receiver, or even both are not aware of AMnet. Typical examples for these, as we call them, *net-centric services* include services like

- Multicast in non-native-IP-multicast environments — One of the AMnet 2.0 service examples automatically and fully transparently pools HTTP music streams to save network and server resources.
- Congestion control for streaming applications — AMnet services can transparently reduce the data-rate of various media streams if congestion occurs.

We believe that such services will promote the deployment of active and programmable network technology by creating a seizable reward for installing active nodes. AMnet 2.0 (for which the described services are either already implemented or about to be implemented soon) can thus now provide gain to a network operator from the first installed node on. We believe that this empowerment of the entity that is required to install and maintain the active nodes is a key element of utmost strategic importance for the actual deployment of AMnet. Benefit from widespread deployment beyond an individual service provider will then automatically occur as a second step. This focus was not present in AMnet 1.0.

According to this extended usage scenario and strategic goal, the AMnet 2.0 architecture has to provide high performance and give control over the achieved security level to the respective domain's administration. Flexibility should primarily be viewed in the light of quick provision of novel services and easy maintenance. This shift in focus required a slight adjustment of the performance-security-flexibility tradeoff as compared to AMnet 1.0:

- AMnet signaling was originally closely bound to the evaluation process that selects a node for execution of a service module. It was also based on evaluation packets that carried the evaluation code (capsule approach). In AMnet 2.0, signaling and evaluation are separated, the capsules are omitted entirely, and the evaluation process is now based on modules drawn from the service module repository (cf. section 2.2). Thus, evaluation and service execution have become very similar. This fact reduces the node's complexity significantly. Together with the complete abandonment of the capsules this increases the overall security considerably.
- AMnet 1.0 employed its own mechanism to grab, mangle, and re-inject packets. With the advent of Linux 2.4 this mechanism became obsolete and could be replaced by a lightweight interface to the standard Netfilter mechanism of Linux. This allows AMnet to be installed most easily on standard Linux nodes. A fact that should also increase trust in the system.

Besides this shift in focus, the new AMnet 2.0 design embodies many practical supplements and extensions of already existing ideas, e.g. a relocation mechanism, improved signaling for off-path node detection, execution environment extensions, etc. These, too, are described in the following sections.

## 1.1    Related Work

Active and programmable networks have been a field of intensive research for quite a while. Therefore, it is impossible to give a brief overview without being forced to arbitrarily choose among the many approaches in this field. We will hence only exemplarily present a few projects to place AMnet in context. Please see, e.g., [18, 2, 3] for a general overview of fundamental concepts of active and programmable networks and [8] for work related to AMnet 1.0.

AMnet's active nodes are Linux based software routers. A similar approach has been pursued by the Click Modular Router project [12]. There, low-level extensions to the regular network protocol stack provide a router environment in which so-called *elements* perform the basic processing steps like packet classification and mangling. Compared to AMnet, which (mostly) runs in the user-space of an unmodified Linux installation, Click's direct hardware access trades security and programming ease against performance. Both projects' common objective, namely to benefit from existing operating system functionality, is also shared by SILK [1], in which a port of Scout [15] replaces the standard Linux protocol stack.

The NodeOS working group [16] gives a general specification of such an interface between the generic operating system and the particular active node functionality, whereas VERA [10] defines an interface between router hardware and software. An example for the use of flexible router hardware is the Dynamically Extensible Router (DER) [13]. Contrary to AMnet, where specialized hardware may optionally supplement the central processing unit of a software router, DER inserts *processing elements* between the *line cards* and the switching fabric.

The rest of the paper is structured as follows: Section 2 explains the main architectural concepts of the AMnode. Section 3 then presents the architectural structure. Section 4 describes our actual implementation of AMnet. Section 5 demonstrates the use of AMnet with a practical example of a service-module. And section 6 finally concludes our presentation with a short outlook on future work.

# 2    Conceptual Overview of AMnet 2.0

AMnet 2.0 consists of two main building blocks, the so-called *service module repository* and the active node, the *AMnode*, itself.

## 2.1    AMnodes as Active Nodes

AMnodes are mainly intended to be positioned within the network, preferably as edge-routers. Being placed on the path data is routed along, they are capable of analyzing and modifying the packets passing by regardless of their actual destination address. Additionally, with AMnet 2.0 AMnodes can also be placed off-path. In that case, AMnet signaling mechanisms provide means to start and control services on these nodes. If necessary, AMnet signaling thus enables multiple AMnodes to cooperate in order to, e.g. redirect traffic from AMnodes on the original path to an appropriate off-path node. Separating the AMnodes that are loaded with all the passing traffic from other AMnodes that have enough resources for computationally demanding services, provides a flexible means for scaling and allows AMnet to handle a broad range of services.

Such services that are flexibly installed within the network are the main guideline of our concept. They can be classified into two major categories: application (or end-system) services and the new net-centric services.

- *Application services* are initiated by an application on an end-system of the network. A typical example is media-transcoding. Here bandwidth-constraints or a request for a specific data-format are signaled from the end-device into the network. Other examples might be network support for peer-to-peer applications or the creation of application-specific overlay networks.
- *Net-centric services* are directly invoked on the AMnodes, e.g. directly during setup by a network administrator or indirectly by other AMnodes. Typical examples are network monitoring, security services like firewalls and intrusion detection systems, protocol-boosters, content-caching or multicast reflector services.

All services within the AMnet scenario are based on *service modules* that contain the actual code for the service. A service can be provided by a single service module alone, or by a combination of several service modules. This possibility to simply chain services modules to create more complex services opens AMnet to a wider audience of programmers whose main skills are scripting languages and not the full details of network programming. This goal is similar to the active pipe approach in [11]. AMnet, however, aims at combining modules within one active node, whereas active pipes are an overlay in the network. Service module programming and the improved chaining of AMnet 2.0 will be described in detail in section 5.

Technically, service modules are small portions of object code that are dynamically linked to the AMnode's *execution environment* upon invocation. Thereby, AMnet combines the high flexibility of on-demand installation of services with native execution speed making it thus feasible to use AMnet beyond mere "in-principle" demonstrations[1]. Besides this capability for native language code, typically C, wrappers for e.g. Java code are also available. We consider this openness to a broad range of languages and programming styles a key feature for a widespread deployment of programmable networking ideas.

Since native code does not run in a sandbox, additional security measures have to be implemented to assure the node's integrity against malicious or erroneous code. AMnet 2.0 introduces a wrapper layer in form of a library between the service module and the node's operating system to impose constraints on the execution of native code. This allows the AMnode to enforce certain restrictions on the service modules [9]. Another key element of AMnet's security strategy is the use of a service module repository.

## 2.2 Service Module Repositories

Service module repositories in AMnet are administratively managed and provide the module descriptions and the modules' object-code, potentially for various platforms. This aspect of the repository is similar to the code caching approach of [4]. Module

---

[1] For a detailed analysis of the achievable system throughput see [6].

descriptions are currently based on a centralized naming scheme only. A more flexible description scheme is under development.

Additionally, service modules can also come in multiple flavors to allow for different resource restrictions or operating conditions. So service modules can be optimized for low memory consumption, high-throughput, or other criteria, like the use of specialized hardware. The AMnet signaling mechanism will select the appropriate flavor of a service module depending on the respective AMnode's capabilities.

Besides the service modules, the service module repository contains service dependent modules for the evaluation and relocation process (as described below). The evaluation and relocation modules have to be service dependent since different services may have different requirements such as the preferred location for the service (close to the receiver or close to the sender) or the number of running service instances on distributed AMnodes (e.g. for the semi reliable multicast service). This mechanism enables the administrator to directly set up preference hierarchies as to which service modules are used within its domain.

The use of a service module repository tackles a major security issue within AMnet. Since the repository is administratively managed and therefore the domain administrator is the only one who can add new services to that repository, only tested and verified code gets executed on the programmable nodes within that domain. Network security protocols (IPSec, SSL) and cryptographic signatures protect the data transfer from the repository to the node from data manipulation and man-in-the-middle attacks.

Service module repositories may be linked to form a network of trust. If a repository cannot satisfy the service request of an AMnode, the search can be extended over all trusted repositories. This implies that service modules can be introduced and updated rather easily at a single insertion point if admitted by the administrative policies. E.g., repository administrators of a given administrative domain could decide to trust a certain software vendor's repository. Code released by this software vendor would then automatically become available on all repositories of that domain. A drawback of this linkage is the danger of quickly spreading faulty code. If employed carelessly, this mechanism of trusted links can bring down a service throughout the network. Care must hence be taken of how trust is established within AMnet, especially if repositories mutually trust one another. Thorough step-by-step testing of new service modules, using AMnet's mechanisms to restrict trusted links by administrative policies, can minimize this threat.

In the following section we present the AMnode's functional architecture in more detail. Implementation specific details are described in section 4.

# 3   Architecture of AMnodes

As already mentioned, AMnodes are intended to be positioned as routers at the edge of the network. Their purpose is to provide services that primarily comprise analysis, conditional forwarding and general modification of the bypassing packets. To achieve this, while observing the already mentioned security requirements, the AMnode architecture forms a three-layered structure containing the following fundamental functional building blocks: packet filtering, signaling and resource monitoring, the execution environment, and support for node selection and service relocation.

At the network-layer a *packet filter* parses the packet stream, transfers packets that match the service modules' filtering rules to the respective modules, and forward nonmatching packets according to the node's routing table. This hook into the operation system's kernel is necessary, since AMnet needs to override the basic networking functionality of a standard router. This building block is fully embedded in the operating system's kernel.

## 3.1 Signaling and Monitoring

The next-higher functions are located in the user space. There the usual Linux APIs apply and simplify the implementation and maintenance of the AMnode.

The *resource monitor* keeps track of the modules' individual resource requirements and initiates countermeasures upon menacing overload. The countermeasure depends on the kind of service module, its priority and the node's administrative policies. Countermeasures include service relocation and forced shutdown of a service. Ideally, relocation should be initiated well before the node's resources are exhausted giving the service enough time to relocate gracefully. In any case, un-cooperating modules must be kept from jeopardizing the node's functional integrity.

The *resource access control* controls the allocation of resources on a per service basis. The security policies are again service-dependent and also stored in the service module repository. The rule set either limits or entirely prohibits the use of certain resources. This access control allows to minimize the threat originating from both, the service modules themselves (e.g. memory exhaustion, malicious operations) and the data processed by the node (e.g. denial-of-service attacks).

The basic *signaling mechanism* provides applications (and thus end-users) with the possibility to request services [17]. It also handles the communication with the service module repositories and allows an AMnode-to-AMnode communication, e.g. for evaluating service requests or to negotiate service relocation.

As mentioned above, the communication between the AMnode and the service module repository is secured either by network or by application security protocols. The same mechanisms can be applied to the inter-AMnode-communication. The communication between end-users and AMnet must include an authentication mechanism to legitimate the request.

Further signaling functionality can then be introduced on a per-service-module level. This is e.g. necessary for protocol-boosters that need to understand the signaling of the respective protocol.

## 3.2 Execution Environment

On top of the AMnode's resource monitor and basic signaling mechanism resides the *execution environment*. Its location in user space with potential additional protection by the mechanisms described in [9] secure the AMnode against malicious service modules.

The *execution environment* hosts the service modules and supplies basic functionality, e.g., for setting and revoking packet filtering rules, access to (parts of) the kernel

API, and to the packets captured by the filter. It thus provides the framework within which service modules are executed. Multiple service modules can potentially be combined to provide more complex services. Technically, the execution environment (EE) is a basic user-space program which the service modules are linked to on demand.

Being located in user-space the EE and thereby the service-modules underly the standard Unix security restrictions for processes. Together with the security mechanisms already mentioned, this ensures that erroneous or malicious service modules can neither manipulate other services nor endanger the AMnode's integrity.

The EE of the first AMnet implementation was limited in various ways, e.g. in the way service modules could be linked together and in the fact that neither branches nor any other kind of packet re-routing, duplication etc. was possible inside the module chain. With the increasing complexity of services implemented in AMnet, a need arose for the flexible combination of multiple modules and the possibility to dynamically change the path on which packets are handed from module to module. An example for such services is a traffic analyzer that dynamically installs new services in the module chain in order to change the behaviour of the service with respect to changing traffic conditions.

Further problems arose with the integration of the resource monitor into the AMnet node architecture. In AMnet 1.0, complex service modules often forked child processes in order to provide proxy or server like functionality. With the introduction of the new resource monitor, forking had to be limited to allow effective resource monitoring. AMnet 2.0 implements several callback handlers that help with common problems usually solved by forking server and proxy child processes and thus avoids many fork operations. A service module can register callback functions, e.g. for timer events as well as for socket and file input-output, allowing a service module consisting of one single process to handle various timers, network sockets, file handles etc. at once. Furthermore, this abolished the need for interprocess communication making the design of complex service modules even simpler and more efficient.

### 3.3   Support for Node Selection and Service Relocation

Three special mechanisms closely related to the service modules themselves, namely evaluation, relocation, and resource management, complete the AMnode architecture:

The *evaluation mechanism* determines on which AMnode a requested service will be started. Primary criteria are administrative policies and the AMnode's current resource availability. If the service request does not have sufficient rights or if the AMnode's available resources fall below the service module's specified threshold, any further evaluation is immediately rejected. Otherwise, a potentially more complex evaluation procedure is started.

The *relocation mechanism* determines to which AMnode an already running service will be relocated upon an overload situation. In most cases, this is very similar to the initial evaluation procedure. For some services however, differing evaluation and relocation mechanisms are better suited, e.g. when a service needs to be quickly started on an in-path AMnode before it is (more time-consumingly) relocated to a better suited AMnode that might be found off-path.

The *resource manager* mediates between the AMnode's resource monitor, the service modules' requirements and the evaluation and relocation mechanisms. Through it, a service can define e.g. its relocation strategy.

In contrast to the fundamental components listed above, these latter functionalities can mostly be built within the service module framework, i.e. they can be coded as service modules that are executed within the node's execution environment. The AMnode itself only needs to provide basic support functionalities for signaling and interaction with the resource monitor. This allows a much greater flexibility than if these mechanism were fixed parts of the AMnet architecture. In most cases however, standard mechanisms can be used. These are realized by ready-made modules available in the repositories.

The following section describes our actual implementation of AMnode. A more detailed documentation together with sample service modules can be obtained with the source code of our implementation [5].

## 4   AMnode Implementation Based on Linux

The central functionality of the AMnode is to provide an interface between the networking core of the host operating system and the service modules running within the AMnode's user-space execution environment. The whole setup is depicted in figure 1.

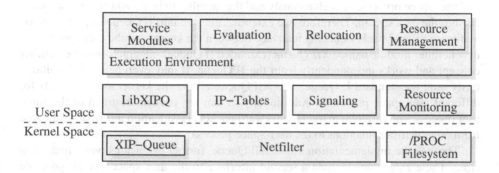

**Fig. 1.** AMnode setup on Linux 2.4

### 4.1   The Use of Netfilter

The first version of the AMnode kernel integration was based on a proprietary hook into the Linux kernel code. The installation of these hooks was available for the 2.0 and 2.2 series of the Linux kernel and required the recompilation of the networking core functions and, hence, a re-installation of the kernel and a system reboot. With the release of the 2.4 Linux kernel, a new flexible interface to the packet processing paths inside the networking core was introduced. This so-called *netfilter interface* forms the basis of functions like packet filtering used e.g. for firewalls and packet modification required e.g. for network address translation (NAT) and IP masquerading. The flexibility of the netfilter interface allowed many third party extensions of the basic routing functionality, e.g., IPsec and QoS.

By using the netfilter interface as the basis of the AMnodes access to the packet processing, it is now possible to use unmodified Linux 2.4 kernels as a basis for the AMnode. This allows the installation and integration of all of the AMnet functionality into a running standard Linux 2.4 system as operated by various ISP's without even the need to reboot the system. AMnet can even be used in combination with other netfilter based extensions[2].

## 4.2   Extended IP Queue (XIPQueue)

The user-level status of the execution environment equips AMnet with a high level of security. On the other hand, this concept of service module based packet processing requires the transfer of network packets from their usual processing space inside the system kernel into the user space where the service modules are located. After the processing in user space, the packets have to be re-injected into the system kernel for further processing by the networking core[3].

The netfilter implementation distributed with standard Linux 2.4 kernels contains an interface module for packet transfer between kernel and user-space called *IPQueue*. Among the many limitations of this IPQueue are the inability to interoperate with several user space processes simultaneously and the inability to inject additional data packets into the data stream. Therefore, we extended the standard IPQueue implementation to a more general packet-queuing interface between the kernel and the user space. This new netfilter module named *XIPQueue* (extended IPQueue) integrates into the netfilter concept and works independently from the IPQueue. It may even be used simultaneously with the standard IPQueue. The XIPQueue allows the kernel to tag packets for different user space processes. Additionally, it can tag packets destined to the same user process with different markers to forward packet classification information derived from other netfilter functions to the user-space process.

Through the implementation of the XIPQueue, further extensions were made possible. These extensions include a second interface to the user space via the proc file system. This interface is used to control and supervise the operation of the XIPQueue and to access queuing specific parameters from user space.

The access from user-space to the XIPQueue netfilter interface is done directly from a user-space application. In order to allow flexible extension and modification of the XIPQueue interface and reuse of the the XIPQueue in other, non-AMnet scenarios, an additional interface layer was implemented. This *LibXIPQ interface library* can be accessed by general user space applications and provides all functions required to access

---

[2] Care must be taken concerning the order of rule installation. This determines e.g. whether the AMnode runs before or after a given other netfilter extension.

[3] For many services (e.g., video-transcoding), the delay caused by the two copy operations is negligible compared to the actual processing time. For high throughput services, we are currently considering improvements that either perform a quick copy based on memory remapping or execute simple operations directly in kernel space.

the XIPQueue netfilter module. Further extensions were made to the *iptables* program which is used to install and maintain netfilter modules in the linux kernel. Extensions to this program are also made via user-space libraries and can be introduced into the existing IPTables installation in a working system.

None of these extensions require the modification or replacement of existing programs or libraries on the system. The XIPQueue netfilter module as well as the LibX-IPQ interface library and the extension library for the iptables program can easily be installed on the running system without the risk of influencing other applications.

## 4.3 The Resource Monitor

The resource monitor is tightly bound to the operating system's kernel. It reads its respective current state and attributes resource usage to the individual services (i.e. modules or module chains). In our implementation this is done via reading Linux' standard proc-filesystem. In order to also include the packet filter status, we provided our packet filtering mechanism with an appropriate interface to the proc-filesystem. Since different service-modules run as different processes, resource usage can be traced via the process identification (PID). Our implementation currently monitors three general parameters: memory, bandwidth, and CPU cycle consumption. Since processing power is difficult to valuate in a general and platform-independent way, we employ an indirect mechanism that has the advantage to yield exactly the required information: Being user-space processes, service modules share CPU time according to their scheduler priorities, i.e. the AMnode can control the modules' individual share of CPU time in the first place. By observing the modules' capability to hold pace with the incoming packet stream the node can determine whether the attributed processing power suffices. A growing queue in the packet filter thus indicates an unsatisfied demand for processing power.

## 4.4 The Execution Environment

A service module (or module chain) is installed and operated by one instance of the user-space based execution environment (EE). The EE is user or script controlled and the service modules are implemented as system shared libraries. So they can be installed on user or script demand, even into an already running EE. Running the modules in the context of their EE allows to control the resource usage of the service modules by observing the behavior of the respective EE user-space process only. Even more important is the fact that a service module in user-space is subject to the same limitations as any other user-space process. Unlike code executed in the kernel, a user-space application usually runs with very limited rights and the risk of critical malfunctions is much smaller when using user-space code than with kernel-space code.

The EE provides (among others) the following fundamental functions:

– *Registration and release of packet filtering rules.* These rules can be composed out of the usual criteria: protocol, address and port range. If necessary, further header fields can be included into the rules.
– *Packet passing, both from the packet filter to the service module and back again into the kernel.* Packets may be modified by the service module. Packets handed

back into the kernel can be marked to be forwarded regularly as indicated by the routing table, to be sent through a specified interface, or to be discarded.

– *Creation of new packets.* Service modules can execute standard system calls unless these calls are prohibited by the security mechanisms described in [9]. This includes the creation of new packets, regardless of their kind. I.e., UDP packets, whole TCP streams, and raw IP packets are equally feasible.

– *Helper Functions.* The EE also provides various helper functions for module chaining, inter-module communication, relocation, and various other tasks common to many service modules.

This concept has major advantages for the service-module developer. A very simple framework forms the core of a service module and helper functions provide additional functionality for all service modules.

## 5   Service Module Example

AMnet draws much of its flexibility from the way modules can be combined out of multiple object code pieces. We will illustrate this mechanism by the following example which sheds light onto AMnet service module programming in general.

### 5.1   Scripts for Module Initialization

The central element of an AMnet service module is a script that controls the module's start-up in the execution environment: it determines which code pieces have to be loaded, where packet filters are to be hooked into the kernel, and which filter rules have to be established.

Figure 2 gives an example of a module that receives all port 80 TCP traffic from (and to) some given server from the kernel's IP pre-routing hook. Note that we normally have to process both directions (i.e. from and to an address/port pair) since content modification typically requires us to modify the acknowledgments, too.

These filter specifications can also be issued or changed with the EE helper functions after module start-up, i.e. from within the module's object code. But the possibility to put such parameters into the start-up script makes service module programming easier for unexperienced programmers. This script based programming idea is further supported by the possibility to specify individual module parameters in the start-up script. In our example we might want to specify a string that can later be used for content-based filtering.

Once the service module is installed in the EE, it receives all packets matching the specified filter rules. In our example, we will receive all port 80 web-traffic from *www.some-server.net* passing through the AMnode. These packets can now be modified in the service module. For inspection of the TCP content, we can, e.g., redirect the data stream to a local socket. This interception mechanism is described in more detail in [5]. It equips our example module with an easy way to inspect and modify TCP content. Information about intercepted TCP connections can e.g. be stored in the EE using the appropriate helper functions for list management. This has the advantage

```
loadmodule tcp {
  file "libtcp_example.so";
  myparameter = "some string";
};

listen {
  protocol "tcp";
  nfhook PREROUTING;
  source {
    name "www.some-server.net";
    port 80;
  };
};

listen {
  protocol "tcp";
  nfhook PREROUTING;
  destination {
    name "www.some-server.net";
    port 80;
  };
};
```

**Fig. 2.** Execution environment script example

that an administrator has direct access to these lists and can easily maintain them from the EE's command line interface. The service module can e.g. now rewrite content as indicated by the parameter in the start-up script. This short example might be used e.g. as a building block for a transparent web-proxy.

## 5.2 Improved Module Chaining

Chaining of service modules was already implemented in AMnet 1.0, but it was limited to a fixed linear combination of multiple modules and mainly used to overcome limitations in filters. For example, in AMnet 1.0 an MPEG2-to-AVI filter could be obtained from chaining an MPEG2-to-MPEG1 filter with an MPEG1-to-AVI filter. Reuse of software components in multiple service modules (e.g., Huffman decoding, IDCT etc.) was possible in principle but required additional communication paths between different service modules.

AMnet 2.0 directly addresses these issues by extending the chaining capabilities and providing several ways of inter-module communication. On top of these new functions lies a powerful chaining scheme that allows to address modules either by function, position in the chain or some unique identifier attached to the modules during its installation [5]. This way, a AMnet 2.0 service module can request additional packets to be forwarded from the kernel to specific service modules. Additionally, service modules can cause packets to be rerouted inside the module chain in order to provide a special packet specific processing. A service module may at any time duplicate packets and schedule duplicated packets for different processing, allowing e.g. to duplicate a data stream and transform the duplicate stream into a different format or to influence the further processing of the duplicate stream by changing destination addresses etc.

Module chaining can be used e.g. to splice together a TCP connection that does no longer need to be intercepted at a local socket. In this, several code pieces are loaded via the start-up script. Packets caught by the packet filter are first handed to a classifier that decides whether this packet needs to be inspected with the help of a local socket or not (cf. figure 3). If it has to be handeled, it is forwarded to the module code described above and handeled as described there. Otherwise it is passed over to another module that does not need to employ a full TCP socket but just rewrites sequence numbers to accommodate for potential modifications in earlier parts of the stream. After having modified a packet, the service module uses yet another helper function to correct the TCP checksum to ensure proper further processing of the packet.

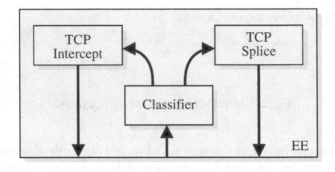

**Fig. 3.** AMnode setup on Linux 2.4

An example for another problem that could not satisfyingly be solved with AMnet 1 was e.g. demultiplexing while splitting a MPEG2 stream with included audio into separate audio and video streams. With the new chaining capabilities of AMnet 2, modules for multiplexing and demultiplexing can easily be implemented at any complexity by changing the routing inside the chain on a per packet basis or by linking several packets paths inside the chain to the same service module.

A service module may also communicate with other service modules or the EE, allowing this module to replace parts of the chain or to change the behaviour of some parts of the chain. This is required e.g. to implement dynamic services that are influenced by the e.g. current traffic situation or specific payloads being observed by a service module. An example for this may be a service detecting a wireless link specific loss pattern on and dynamically loading a service module to apply forward error correction on the affected link or apply protocols especially suited to handle the current situation on the link.

# 6 Conclusion and Future Work

In this paper we gave an overview over the current state of the AMnet project. We described the basic concepts of AMnet and the architecture of its active node, the AMnode. The three-layered approach of kernel-level packet-filter, fundamental node support (signaling and resource monitoring), and execution environment was motivated, and the interaction of EE and service modules illustrated by a short example.

Our future work aims at increasing the number of available service modules, thus producing building blocks for individual services and applications. We believe that the experience with productive uses of the AMnet architecture can provide valuable experiences that help us in further extending AMnet towards a practical programmable network infrastructure.

**Acknowledgments**    The work described here has been performed in the framework of the FlexiNet project which is funded under grant number 01AK019E by the Bundesministerium für Bildung und Forschung. The authors would like to thank the reviewers of the paper and its shepherd for the helpful comments.

# References

[1] Andy Bavier, Thiemo Voigt, Mike Wawrzoniak, Larry Peterson, and Per Gunningberg. SILK: Scout paths in the Linux kernel. Technical Report 2002-009, Uppsala Universitet, February 2002.

[2] Kenneth L. Calvert, Samrat Bhattacharjee, Ellen Zegura, and James Sterbenz. Directions in active networks. *IEEE Communications Magazine*, 36(10):72–78, October 1998.

[3] Andrew T. Campbell, Herman G. De Meer, Michael E. Kounavis, Kazuho Miki, John B. Vicente, and Daniel Villela. A survey of programmable networks. *ACM SIGCOMM Computer Communication Review*, 29(2), April 1999.

[4] Dan Decasper and Bernhard Plattner. DAN: Distributed code caching for active networks. In *Proceedings of INFOCOM'98*, San Francisco, CA, April 1998.

[5] Thomas Fuhrmann, Till Harbaum, Marcus Schöller, and Martina Zitterbart. AMnet 2.0 source code distribution. Available from http://www.flexinet.de.

[6] Till Harbaum. *Rekonfigurierbare Routerhardware für adaptive Dienstplattformen*. PhD thesis, Insitut für Telematik, Universität Karlsruhe, 2002.

[7] Till Harbaum, Anke Speer, Ralph Wittmann, and Martina Zitterbart. AMnet: Efficient heterogeneous group communication through rapid service creation. In *Proceedings of the 2nd International Workshop on Active Middelware Services (AMS'00)*, Pittsburgh, Pennsylvania, August 2000.

[8] Till Harbaum, Anke Speer, Ralph Wittmann, and Martina Zitterbart. Providing heterogeneous multicast services with AMnet. *Journal of Communications and Networks*, 3(1):46 – 55, March 2001.

[9] Andreas Hess, Marcus Schöller, Günther Schäfer, Adam Wolisz, and Martina Zitterbart. A dynamic and flexible access control and resource monitoring mechanism for active nodes. In *Proceedings of the 5th International Conference on Open Architectures and Network Programming (OPENARCH'02)*, New York, NY, June 2002.

[10] Scott Karlin and Larry Peterson. VERA: An extensible router architecture. *Computer Networks*, 38(3):277–293, February 2002.

[11] Ralph Keller, Jeyashankher Ramamirtham, Tilman Wolf, and Bernhard Plattner. Active pipes: Service composition for programmable networks. In *Proceedings of Milcom*, Washington DC, October 2001.

[12] Eddie Kohler, Robert Morris, Benjie Chen, John Jannotti, and M. Frans Kaashoek. The click modular router. *ACM Transactions on Computer Systems*, 18(3):263–297, August 2000.

[13] Fred Kuhns, John DeHart, Anshul Kantawala, Ralph Keller, John Lockwood, Prashanth Pappu, David Richards, David Taylor, Jyoti Parwatikar, Ed Spitznagel, Jon Turner, and Ken Wong. Design of a high performance dynamically extensible router. In *Proceedings of the DARPA Active Networks Conference and Exposition (DANCE)*, San Francisco, May 2002.

[14] Bernard Metzler, Till Harbaum, Ralph Wittmann, and Martina Zitterbart. AMnet: Heterogeneous multicast services based on active networking. In *Proceedings of the 2nd Workshop on Open Architectures and Network Programming (OPENARCH'99)*, New York, NY, USA, March 1999.

[15] David Mosberger. *Scout: A Path-based Operating System*. PhD thesis, Department of Computer Science, University of Arizona, July 1997.

[16] Larry Peterson. *NodeOS Interface Specification*. Active Networks NodeOS Working Group, Department of Computer Science, Princeton, January 2002.

[17] Anke Speer, Marcus Schöller, Thomas Fuhrmann, and Martina Zitterbart. Aspects of AMnet signaling. In *Proceedings of the Second International Networking Conference*, pages 1214–1220, Pisa, Italy, March 2002.

[18] David L. Tennenhouse, Jonathan M. Smith, W. David Sincoskie, David J. Wetherall, and Gary J. Minden. A survey of active network research. *IEEE Communications Magazine*, 25(1):80–86, January 1997.

# Design and Implementation of a Python-Based Active Network Platform for Network Management and Control

Florian Baumgartner[1], Torsten Braun[2], and Bharat Bhargava[1]

[1] Department of Computer Sciences and Center for Education and Research in Information Assurance and Security (CERIAS), Purdue University, West Lafayette, IN 47907, USA
`baumgart|bb@cs.purdue.edu`
[2] Institute of Computer Science and Applied Mathematics
University of Berne, Bern, Switzerland
`braun@iam.unibe.ch`

**Abstract.** Active networks can provide lightweight solutions for network management-related tasks. Specific requirements for these tasks have to be met, while at the same time several issues crucial for active networks can be solved rather easily. A system addressing especially network management was developed and implemented. It provides a flexible environment for rapid development using the platform-independent programming language Python, and also supports platform dependent native code. By allowing to add new functions to network devices it improves the performance of Internet routers, and simplifies the introduction and maintenance of new services.

To show the capabilities of the approach, two different quality of service related applications, that is a simple multicast algorithm and an approach to automatically set up tunnels, have been implemented. The evaluation of these services shows the advantages of the architecture, and its benefits for the task of network and quality of service management.

## 1 Introduction

The limitations of current management mechanisms to provide an adaptable system to configure and control services within the network of Internet Service Providers (ISP) enforce the development of new strategies for network management.

In addition to a distributed resource management, mechanisms are needed to add new services dynamically to the network. For competing ISPs the time needed to provide a new service to their customers is essential. This is why an architecture allowing the dynamic and quick establishment of new Internet services is important.

The most obvious way to provide such flexible systems is the exchange of program code between network nodes. The program code is executed by a network device, reconfiguring the router, collecting information or sending data and program code to other devices. Networks providing such mechanisms are called active networks. Such an approach is even more powerful since the differences between classical network management and signaling are becoming smaller. Active networks can not only be used to provide a high degree of control over the network but can easily be used to implement

J. Sterbenz et al. (Eds.): IWAN 2002, LNCS 2546, pp. 177–190, 2002.

lightweight signaling mechanisms, needed for the control of large networks. There are several types of active networks.

The capsule approach, as presented by Tennenhouse [19], uses packets consisting of a short piece of code and of additional payload. An active router receives the capsule and executes the code. This code may simply be used to route the capsule to the destination or to configure the active router on the path. Another approach focuses more on a programmable network [7] and proposes a more moderate active networking model. Active network mechanisms are used for signaling, configuration and monitoring purposes [11].

The execution of code on network devices causes several problems. Especially with the capsule approach security and performance are still open issues, but even within moderate active networks, the security issue is not solved yet. Murphy [14] shows the security problems and proposes an architecture for active networks. A different approach to improve security was proposed by Brunner [6], who suggests the creation of Virtual Active Networks. Similar to virtual private networks (VPNs), a customer can "rent" such a virtual active network. Within its active network, the customer (e.g. a company) can exploit AN technology but has no access to the other virtual networks.

The goal of the approach presented within this paper is to provide an easy to use, lightweight mechanism with a focus on management related tasks. In contrast to systems with a broader approach, this reduces the complexity of several typical active networking problems, like security and performance. On the other hand, such a system requires a more direct interaction between the IP router and the active components. Instead of providing a distributed application platform, which mainly uses the IP network as a transport medium, the system has to change and modify router functions, add new components, or has to be able to influence packet processing within the router core. This requires lightweight but powerful and flexible interfaces to native code and functions as well as a high level programming language to support the rapid development of configuration scripts.

## 2    The Python-Based Active Router

Various languages have been discussed and used for the implementation of active networking systems [18]. While PLAN [13] is based on a functional language like Caml [8], others, like the Active Network Transport System (ANTS) [22], are based on Java. In contrast to those, the active networking architecture proposed here was developed especially for the purpose of network management and therefore uses the Python [17] language, which provides certain advantages for that kind of application.

### Python as a Language for Active Network Systems

Python meets the properties of most modern, interpreted languages, used for active networks. It provides portable bytecode, which allows an execution on different platforms; restricted execution environments to keep active packets from damaging the environment; and threads.

One of the strengths of Python, which makes it especially useful for configuration related tasks, is its extensibility. In contrast to Java with its native interface (JNI) [7] or other languages, the integration of native code modules has been a central aspect of the Python programming language from the beginning. This is also the reason why many applications use Python as a configuration and control front-end and rely on native code for time consuming tasks. The functional separation used by these programs matches perfectly the situation we face on an active IP router, with high speed packet processing in the kernel and an interpreted language on the control plane.

Another argument, which led to the use of Python instead of Java or Caml, was the type of programs we expect to be sent over the network. Similar to system administrators, automating certain tasks, network administrators have to be able to write short configuration scripts and send them through the network. Since these scripts will highly depend on the current network state and on a specific task, the capacity of a language to support rapid development is important. Python provides high-level data structures and dynamic types, which support the rapid development of configuration scripts. Since more generic systems face security and performance problems, Alexander [2] recommends different, or even contrary properties for active networking languages. As will be shown in the next section, the specialization of the PyBAR system reduces this problems, and therefore allows to benefit from Python's prototyping capabilities.

Due to its high level data structures and dynamic types, Python programs are short. Their source code is usually three to five times shorter than comparable Java sources [20] and also the bytecode used by the PyBAR consumes significantly less memory. Therefore it is possible, similar to special purpose languages like Sprocket or Spanner [18], to transmit reasonable programs within a single Ethernet packet. This simplifies the transmission of active packets crucially, since no fragmentation/reassembly and reliability mechanisms are necessary.

A general comparison between Python and other languages is presented in [20]. For a detailed performance evaluation of Python and Java see [12]. The Python based active router (PyBAR) system has been implemented for Linux and for Virtual Routers [3]. A Virtual Router is an IP router emulator, which can be combined with real networks. This allows real routers and hosts to be integrated into an emulated network.

## 2.1    The PyBAR Architecture

The design of the active network platform tries to separate the active components as far as possible from the conventional router functionalities, but provides access to routers internal components, like traffic conditioning systems, packet filters and routing. Such a separation ensures portability and also allows an easier integration of existing devices. Figure 1 shows the general architecture of the system.

The PyBAR architecture is based on the standard Python virtual machine and can be connected to several network nodes (e.g. routers). In addition to a rather thin NodeOS (platform adaptor) layer written in C++, the system consists of a set of native or interpreted library and extension modules and a central core written in Python to execute received code. Received packets are forwarded either to a specific service handler, provided by an extension module, or are processed by the core.

The NodeOS provides communication facilities for the PyBAR core and the extension modules. Instead of running merely on top of the IP router, the NodeOS provides several interfaces to the routers, using Python's capability to use native code. This allows the addition of new functions like traffic conditioning, encapsulation or monitoring components directly to the IP routers kernel.

**Security** Since the system is only used by network administrators and their management tools, security issues are less complex to solve than in more generic environments [14][11]. The question of whether a platform is trustworthy or whether a secure bootstrap mode [2] is supported, is of less importance, since the devices are owned and controlled by the same network provider.

The basic mechanism used by the PyBAR platform is a public key infrastructure, which can be used to sign and encrypt active packets. For that purpose the system provides a set of encryption mechanisms, which are provided by the PyRSA extension module allowing the flexible use of high speed cryptographic algorithms. Cryptographic mechanisms require high processing speed, and the module therefore was implemented in C. This solution provides high usability with good performance.

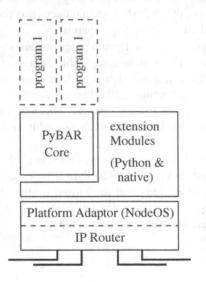

**Fig. 1.** Basic architecture of the PyBAR system.

Since active packets can be modified during transmission, authentication for active packets is complicated [14]. A PyBAR active packet consists usually of several parts (code blocks), which allows sensitive (e.g. executable) code blocks to be static and uses other code blocks for the volatile data [11]. Additionally, within the PyBAR system active packets can be forwarded within a specific Differentiated Service class. Access to this Differentiated Services class can be limited to devices operated by the network provider itself. By setting up filter mechanisms at border routers, a modification of active packets can be prevented. This approach is similar to the Virtual Active Networks [6], but uses a Differentiated Services class instead of a VPN-like mechanism to control the access to the active network.

**Code Transport** The advantage of active networking regarding network management is that code can be transported along the path with normal data traffic. Therefore, programs can become active at nodes requiring configuration without keeping central structures like topology databases.

– A packet has one of the router addresses and is forwarded to the PyBAR for further processing. This mechanism is especially useful to address specific end systems or to send an active packet directly to a special active router.

- The Router Alert option is an IP option, indicating that a router should treat the packet in a special way and can be used to trigger active packet processing [18]. The Router Alert option was introduced in conjunction with the RSVP protocol and has the purpose to send a packet to a specific destination and trigger certain functionalities in the routers along a certain path.
- Some IP routers process packets containing IP options like Router Alert more slowly than normal packets. To prevent a delay of active packets, a special Differentiated Services Code Point [4] value can be used instead of the Router Alert option. Active routers forward these packets to the active components while conventional devices will simply ignore this DSCP value. Besides improved performance this has other advantage. Since packets with such a DSCP can be handled preferentially by Differentiated Services routers, the loss of active packets can be ignored, and limiting the use of that DSCP to devices of the network provider reduces security risks. Of course multiple DSCPs may be used as well. As an example, there might be a DSCP value for active signaling packets and another for active packets carrying data.

**Extension/Library Modules** Since the hard coded commands provided by the PyBAR cover only very fundamental tasks, and the availability of functions may depend on the platform, more complex issues have to be covered by extension modules.

Extension modules can be provided by native code or by Python. Since the integration of native code and libraries is a fundamental mechanisms in Python, already existing libraries can be added without much overhead, similar to the PyRSA module. Of course, an appropriate module for the current platform has to be available, but since there is no difference between calling functions provided by a native module, or using an interpreted Python module, a platform-independent Python version of a module may be provided as a default.

Any of these modules can be added, replaced, or removed dynamically by active network mechanisms. A high-level, interpreted extension module to provide a uniform interface to set up resources will be presented in Section 2.2.

**Packet Processing and Code Execution** The PyBAR core is responsible for the treatment of received packets containing code. The active service id field in the packet header signals the platform adaptor which active packet type is received. While packets containing executable code are forwarded to the core, the processing of other packets is left to active service handlers. Comparable to router plug-ins [9], modules can be installed to process certain packets, identified by an active service id.

The processing of executable packets is much slower and time consuming than the simple forwarding of a packet's payload. Furthermore, the treatment of such streams can be accomplished completely by the NodeOS and specialized extension modules. If the service handler is provided by native code, the complete processing of such a packet is "Python-free" and therefore reasonably fast.

If a packet contains a program to be executed, the entire packet has to pass consistency and security checks. While in the first step basic properties of the packet are controlled, the second one covers digital signatures and encryption related issues. After the packet has passed the consistency and authentication tests, an execution environment is initialized and the code within the packet is executed.

The core is Python-based and mainly a kind of framework, which can be adapted to different needs and to provide very different mechanisms. Typical functions provided by the core are:

- Monitoring of a running code. This monitoring provides no absolute security, but merely limits the damage caused by program errors.
- A small central database or stack is provided to allow capsules to store and exchange data.
- Functions to install and replace modules within the library are provided.

In the description of the Python language the "high level" of the Python language was mentioned. Even if some of the tasks, like cryptography, require rather complex algorithms, the core itself is very simple and uses only some high-level data structures and functions provided by those modules.

**Packet Format** Packets, which contain executable code, can use the same basic header as packets, which only have to be processed by some active component. To avoid as much overhead as possible, this header is much simpler and shorter than an ANEP header [1]. This increases the performance of packet processing by service handlers, and leaves the processing of more sophisticated header fields, which are required for executable packets, to the Python-based core.

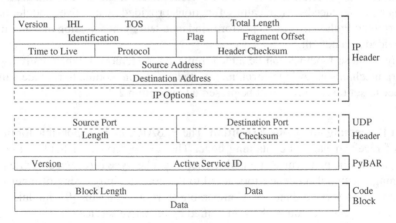

**Fig. 2.** PyBAR IP/UDP payload and a PyBAR code block. Several code blocks might be attached to a packet.

The minimum required PyBAR header is exactly four bytes long (see Figure 2) and can be encapsulated either in raw IP or in UDP packets. It only contains an eight-bit wide field with version information and a 24-bit wide active service identifier. The latter allows to notify which module of the PyBAR system shall process the packet's payload. Since this header is very small and simple, a stream packet has to be checked only for the active service id and can be forwarded to the corresponding module responsible for this type of service.

The special active service identifier is used if the rest of the packet contains executable code. If the active service identifier signals an "active" payload the rest of the payload has to consist of at least one code block. A code block consists of a 16 bit wide length field and the code block's content only. Multiple code blocks may be contained within a packet. The PyBAR system expects the first code block to contain executable code. Besides the program, this code block also contains a signature and other information. However the payload format of the first code block is completely handled by the PyBAR system itself and can therefore be adjusted easily. The other code blocks are not processed and it is left to the code in the first code block to handle them. The platform adaptor provides mechanisms allowing to access and manipulate the other code blocks in the packet. This includes the execution of code blocks and their installation as PyBAR library modules.

## 2.2  Differentiated Services Support

A central problem within the Internet is the lack of a homogeneous configuration interface to network devices. Dependent on the vendor, very different interfaces are provided. Even more problematic than the interfaces are the fundamental design differences of devices [21].

A good example for such different design concepts are traffic conditioning components, which are rather different on various platforms. Therefore, a simple mapping of configuration commands might not be feasible. The Linux traffic control system uses a concept of nested boxes. Each box is a traffic conditioning component, which can contain other components. Other implementations are more list oriented or use a graph-like layout of their traffic conditioning systems like the Virtual Router does.

These different concepts prohibit the use of low-level configuration commands within a network. In contrast to providing a general application programming interface [15] the PyBAR simply offers different commands for different types of platforms. Since a capsule might be executed on multiple hosts and a distinction among the different command sets is not feasible, those low-level commands are usually not used but library modules are installed on the system in advance. A module providing methods to set up and maintain Differentiated Services on a router can map a command set to the low-level configuration scripts. Several high-level functions can be provided by such a module:

- Functions to initialize and configure the basic system have to be provided. Some kind of init() function may set up the complete set of queues, schedulers and classifiers needed to provide Differentiated Services. Parameters of the init function may define whether an ingress, egress or intermediate router has to be set up.
- The marker mechanisms have to be configured to mark packets with different DSCPs. Functions to add or remove flow descriptions at the ingress routers have to be provided. The mechanisms to apply such flow descriptions to the router may vary significantly. Therefore, a high level interface would be beneficial.

– Since the Differentiated Services concept requires a separate handling of the different traffic classes, each traffic class has to be configured for a certain share of the link bandwidth. Therefore, also these parameters are important for a general interface.

Obviously, such a description can never meet all aspects of Differentiated Services or of any other Quality of Service- providing mechanism. Therefore the PyBAR does not even try to provide such a general interface or even a multi-platform module providing such a functionality.

Therefore, a module integrating DiffServ configurations for Virtual Routers and for Linux was implemented, providing a convenient small set of commands as listed in Table 1. The module is written in Python and can rather easily be extended to provide more control and advanced features. This is important, because the set of commands provided by an appropriate Differentiated Services module depends on the services an Internet Service Provider wants to provide and control.

| init() | sets up the complete traffic conditioning components required for DiffServ, with a an appropriate scheduler, EF and AF queues, token bucket filters |
|---|---|
| setClassShares(...) | configures the bandwidth shares for the different traffic types |
| mark(...) | configures the Differentiated Services marker to mark specific flows with certain DSCPs |
| unmark(..) | removes a marker rule |

Table 1. Commands provided by the DiffServ module for the configuration of Differentiated Services resources.

This explains why a lightweight mechanism to install modules on different platforms and to adapt extension modules for new purposes is much more important than the attempt to provide a really generic interface. The mechanisms to install such modules are provided by the PyBAR platform. An active packet may transport and install code within each suitable node of a network, on a single device only, or on any machine along a certain path. Even the attachment of multiple code objects is possible, whereas each code object is to be installed on the matching router hardware.

## 3   Adding Active Services to a Network

The active service id field in the PyBAR header provides an easy mechanism to add various packet treatments. Since the processing can completely be provided by native code, the required processing power stays reasonably low, even for more complex algorithms. A possible application for such active services can be the support for video applications or the support for management related tasks, like a framework to establish tunnels within a network, or simple multicast mechanism for small groups.

## 3.1   Active Tunnel Establishment

IP tunnels are an enabling technology for the application of new services and network management. Even if the basic mechanism – an encapsulation of a packet into another packet (IPIP) as proposed by [16] – is simple, more complex functions may be realized:

- The encryption and decryption of packets at the tunnel start and end points can provide end-to-end security. This way flows can be transmitted securely without the danger that an untrustworthy service provider might eavesdrop.
- Since all packets transmitted through a tunnel get a new header at the tunnel start and the tunnel end point addresses, traffic conditioning mechanisms can be applied quite easily for all packets in the tunnel, even if the encapsulated packets have different source and destination addresses.
- Tunnels allow transport of signals transparently through a network and keep them from triggering mechanisms within an ISP (e.g. an RSVP reservation setup).

For the establishment of tunnels, appropriate start and end points are required, if certain special services like encryption have to be provided. But even for a simple IP in IP encapsulation, an end point needs to be capable of handling the decapsulation of packets.

Since the configuration of a tunnel is usually sender-driven, the general problem is to detect an appropriate end point. When establishing a tunnel between border routers of an ISP, the ingress border router does not usually know the address of the egress border router. In contrast to other approaches using active networks to establish tunnels [13], the focus here is not to provide tunneling mechanisms by active code itself, but on finding appropriate end points and setting up encapsulation mechanisms within the IP router. The active components are only involved for the instantiation of the tunnel, and leave the encapsulation of packets to the IP router.

An active network allows triggering the establishment of tunnels automatically without the need for of any additional protocol. Since an active packet allows the definition of a kind of "search pattern" for an end point and can simply be sent downstream towards the destination, the establishment of a tunnel can be simplified and also the exchange of keys might be accomplished.

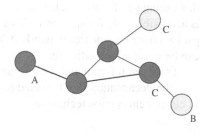

**Fig. 3.** A network with routers capable to tunnel packets (grey) and routers without that mechanisms (white).

Figure 3 shows a small network with a set of grey routers capable of handling tunnels and white end systems without this support. Node A wants a tunnel to be established as close as possible to point B. To find out an appropriate end node, an active packet addressed to node B can be injected into the network. The packet will pass the nodes along the path and check whether the node is an appropriate tunnel end point or not. Similar to the traceroute program reporting each passed router, the active packet can report each possible candidate for a tunnel end point to node A. If simple IPIP tunnels, not requiring a specific setup of the tunnel end point,

```
class DiscoverEP(ARpacket):
    def __init__(self,acpkt):
        # get a list of router properties/services
        c=pad.getCaps()
        # if IPIP available, extract information from
        # code block and send feedback packet
        if c.count('IPIP'):
            src_info=cPickle.loads(acpkt.cb(1))
            # generate and send feedback packet
            p=pad.UDPPacket()
            p.source=pad.hostip
            p.dest=src_info['tunnel_start']
            p.destport=src_info['portnumber']
            p.payload=cPickle.dumps({'service':'IPIP',
                'tunnel_end':pad.hostip, 'time':pad.time})
            p.send()
        # forward original active packet
        acpkt.send()
        return
```

**Table 2.** Active Packet code to discover a device, able to handle IPIP tunnel endpoints.

are used, node A simply uses the most appropriate candidate as the end point and sends the encapsulated packets to this address.

Table 2 shows a the active packet's Python code to check whether a device can handle an IPIP tunnel. The whole packet consists of the first code block containing this executable program and a second code block containing information about the tunnel start point (src_info) like the address and the port number to which the feedback packet has to be sent.

To ensure that such a configuration packet was sent by an authorized network node, this exchange of active packets takes place within a Differentiated Services class, which is accessible only for specific nodes and prevents a transmission of active configuration packets from outside that network. Additionally, such a packet carries a signature, which can be checked periodically.

This search for an end point is not only determined by the simple capability to provide the decapsulation of tunneled packets, but may also be used to find nodes capable of certain encryption techniques.

## 3.2   A Simple Active Multicast Service

Another example for an application using active services is the simple active multicast mechanism, which is useful to support multicast for small groups. The active service id signals the existence of additional IP addresses in the packet's payload.

A similar but not active multicast service for small groups (xcast) has been proposed by Boivie et al. [5]. An active approach to distribute multicast packets using functional language elements of the PLAN system is described by Hicks [10]. In contrast to the approach using PLAN, the algorithm presented here conforms better to the current xcast approach. The main difference between the approach presented here and xcast, is the way an xcast packet is signaled. While xcast uses a dedicated destination address, this algorithm uses the active service id to trigger the special packet treatment. Especially if a native service handler is provided, an active router running the PyBAR system can easily be updated to provide a xcast mechanism at reasonable speed.

Figure 4 illustrates the general concept. A server has to send identical data to multiple addresses. Instead of transmitting one packet via unicast to each destination, one packet is sent, containing multiple addresses. An active router on the path detects the packet due to a special DSCP values, checks the active service id, and ex-

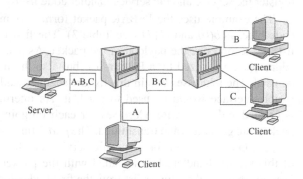

**Fig. 4.** Processing of a packet with multiple addresses.

tracts the list of addresses. If all addresses are routed over the same interface, the original packet is forwarded, otherwise the packet is converted into several packets, each carrying the addresses of the clients routed over the same interface. The resources saved in comparison to unicast are evident.

**Table 3.** The complete code of the service handler providing a simple multicast service.

```
class MiniMulticast(ARservicehandler):
    def forward(self,pkt):
        # extract address list from first code block
        alist=cPickle.loads(pkt.cb(0))
        ifcs={}
        # scan address list and create interface/address-list structure
        for i in alist:
            interface=pad.queryRoute(i)['if']
            if not ifcs.has_key(interface):
                ifcs[interface]=[]
            ifcs[interface].append(i)
        # scan interface/address-list structure and send packets
        for i in ifcs.keys():
            p=pad.Packet()
            p.cb(0)=cPickle.dumps(ifcs[i])
            p.cb(1)=pkt.cb(1)
            p.send({'dest':ifcs[i][0], 'ptype':pkt})
        return
```

Two extension modules providing the appropriate service handler have been implemented. While the first one uses Python only, the second one is platform dependent using C++. Table 3 lists the complete Python code required for the service handler. An active packet for installing this service handler will usually consist of a short script to register the service and the service handler code itself.

The example uses the PyBAR packet format. A multicast packet consists of two code blocks cb(0) and cb(1) (see Table 3). The first code block contains the address list, the second one the payload of the packet. As a simplification, in the example the addresses are realized by a Python list, but also the more generic xcast format could be used. The addresses within this list are scanned and for each address it is checked, which interface would be used to reach it. As intermediate results, a data structure is created and filled with addresses for each outgoing interface. In a final step new packets are generated and transmitted. The *pad.\** functions are provided by the platform adapter. Once a packet for a specific service handler arrives, the *forward()* function of this service handler class is called with the packet as a parameter. The *forward()* function extracts the address list from the first code block of the packet (*cPickle*), scans the addresses, sets up the dictionary and finally creates and sends the new packets. The destination address of a new packet for a certain interface is chosen from the list of addresses routed over that interface.

Table 4 shows the performance[1] of the Python and the C++ based service handlers for different numbers of addresses. The bandwidth calculation was based on the execution speed of the Python code and an average packet size of 1000 bytes. For the output bandwidth it was assumed that each packet is split up completely, and each address has to be routed over a different interface.

| nr.of addresses | time [ms] | packet rate (in) | packet rate (out) |
|---|---|---|---|
| Python module | | | |
| 4 | 1 | 1000 | 4000 |
| 8 | 1.7 | 580 | 4640 |
| 16 | 2.2 | 454 | 7264 |
| Native extension module (C++) | | | |
| 4 | 0.014 | > 10000 | - |
| 8 | 0.026 | > 10000 | - |
| 16 | 0.05 | > 10000 | - |

**Table 4.** Performance of the simple multicast service provided by service handlers written in Python and in C++.

The packet rates reached by the native extension module could not have been measured exactly, but they were more than sufficient to process incoming packets at a full link bandwidth of 100 Mbps. The performance of the Python example can not compete with the implementation using the native code but could being improved by using a more sophisticated format to store the destination IP addresses.

---

[1] Measured on a Linux 400 MHz Pentium II

# 4   Summary

The Python-based Active Router (PyBAR) system is a lightweight active networking platform focusing on the tasks of network management and configuration. The system provides flexible access to router resources and forwarding mechanisms, which is necessary to provide quality of service or to configure devices. Both platform-independent and native extension/library modules are supported, and allow, if a native module is available, to perform packet processing without the limitations of interpreted code.

The system provides basic security by using standard encryption and authentication mechanisms. Additionally active packets can be transported within a specific Differentiated Services class, which prevents unauthorized injection of active packets, and also supports a preferential treatment of active packets.

The support for rapid development, provided by Python, is pursued by ongoing work to combine Python and Java with the goal to be capable to use prototypical Python programs from within Java applications, and directly compile Python programs to Java bytecode. These prototyping capabilities allow a rapid development of configuration scripts, which is important for the usability of the system. Additionally, the high-level character of the language reduces the program size and improves the system performance.

The capabilities of the approach were demonstrated by two classical active network applications. Using the system for the establishment of tunnels, illustrates how to automate and distribute configuration tasks with PyBAR. Active packets can be used to set up tunnel endpoints according to certain specifications. The second example is a multicast mechanism for small groups, which shows the advantages of installing new services on the network, and serves for a short performance comparison of a platform independent Python approach and an implementation using native code.

# 5   Conclusions

Seamless integration of native code modules within the PyBAR provides great advantages for the task of network and quality of service management. Resources within the network devices can be accessed and native code can be used more easily than using languages like Java. This allows the easy installation of new traffic conditioning components. The capacity to provide packet processing by the native code only allows to implement services requiring high performance. This encourages the implementation of native mechanisms for the PyBAR, instead of the underlying platform itself, increasing the flexibility and portability of new services.

# Acknowledgment

The work described in this paper is a part of the work done at the University of Bern in the project 'Quality of Service Support for the Internet Based on Intelligent Network Elements' funded by the Swiss National Science Foundation (project no 2100-055789.98/1) and the SNF R'Equip project no. 2160-53299.98/1. This work was also supported in part by the NSF grant EIA 0103676.

# References

[1] D.S. Alexander. Active network encapsulation protocol. CIS, University of Pennsylvania, http://www.cis.upenn.edu/ switchware/ANEP/, August 2002.

[2] D.S. Alexander, W.A. Arbaugh, A.D. Keromytis, and J.M. Smith. A secure active network environment realization in switchware. *IEEE Network*, 12(3):37–45, 1998.

[3] F. Baumgartner and T. Braun. Virtual routers: A novel approach for qos performance evaluation. In Crowcroft e. al., editor, *Quality of Future Internet Services*, LNCS, pages 336–347. Springer, 2000. ISBN 3-540-41076-7.

[4] S. Blake, D. Black, M. Carlson, E. Davies, Z. Wang, and W. Weis. An architecture for differentiated services. Internet Standard RFC 2475, December 1998.

[5] R. Boivie, N. Feldman, Y. Imai, W. Livens, D. Ooms, and O. Paridans. Explicit multicast (xcast) basic specification. Internet Draft draft-ooms-xcast-basic-spec-03.txt, June 2002. work in progress.

[6] M. Brunner and R. Stadler. Virtual active networks - safe and flexible environments for customer-managed services. In R. Stadler and B. Stiller, editors, *Active Technologies for Network and Service Management*. Springer, 1999. ISBN 3-540-66598-6.

[7] Ken Calvert, Samrat Bhatacharjes, Ellen Zegua, and J.P.G. Sterbenz. Directions in active networks. *IEEE Communications*, 36(10):72–78, October 1998.

[8] The Caml language, INRIA, France, http://caml.inria.fr, June 2002.

[9] D. Decasper, Z. Dittia, G. Parulkar, and B. Plattner. Router plugins: A software architecture for next generation routers. In *Proceedings of the SIGCOMM Conference*, 1998.

[10] M. Hicks, P. Kakkar, T. Moore, C.A. Gunter, and Scott Nettles. Network programming using plan. In B. Belkhouche and L. Cardelli, editors, *Proceedings of the ICCL Workshop, Chicago*, LNCS. Springer, 1998. ISBN 3-540-66673-7.

[11] A.W. Jackson, J.P.G. Sterbenz, and R.R. Condell, M.N. Hain. Active monitoring and control: The sencomm architecture and implementation. In *Proceedings of the DARPA Active Networks Conference and Exposition (DANCE)*, pages 379–393. DARPA, 2002.

[12] G. Lefkowitz. A subjective analysis of two high level, object oriented languages. http://www.python.org/doc/Comparisons.html, April 2000.

[13] J.T. Moore, M. Hicks, and S. Nettles. Chunks in plan: Language support for programs as packets. In *Proceedings of 37th Annual Allerton Conference on Communication, Control, and Computing*, 1999.

[14] Sandy Murphy. Security architecture for active nets. http://www.dcs.uky.edu/~calvert/arch-docs.html, May 2001. AN Security Working Group.

[15] Proposed ieee standard for application programming interfaces for networks. http://www-.ieee-pin.org.

[16] C. Perkins. IP encapsulation within IP. Internet Standard RFC 2003, October 1996.

[17] Python language website. http://www.python.org, August 2002.

[18] B. Schwartz, A.W. Jackson, W.T. Strayer, W. Zhou, R.D. Rockwell, and C. Partbridge. Smart packets: applying active networks to network management. *ACM Transactions on Computer Systems*, 18(1):67–88, 2000.

[19] D. Tennenhouse et al. A survey of active network research. *IEEE Communications Magazine*, January 1997.

[20] G. van Rossum. Comparing python to other languages. http://www.python.org/doc/essays/-comparisons.html, August 2002.

[21] W. Wang and J. Biswas. Standardizing programming interfaces for tomorrow's telecommunications network. *IEEE Standard Bearer*, 12(2), April 1998.

[22] D. Wetherall, J. Guttag, and D. Tenenhouse. Ants: A toolkit for building and dynamically deploying network protocol. In *IEEE Openarch*, April 1998.

# Designing Service-Specific Execution Environments

Mary Bond, James Griffioen, Chetan Singh Dhillon, and Kenneth L. Calvert*

Laboratory for Advanced Networking
University of Kentucky
Lexington, KY 40506

**Abstract.** Past work on the *active network architectural framework* has focused on the NodeOS and Execution Environment (EE), which offer rather standard programming environments. As a result, *Active Applications (AAs)* must build the desired service from the ground up. Ideally, active applications could be built on higher-level *active services* that could themselves be "programmed". In this paper, we examine the issues involved in the design and implementation of higher-level active network services. We describe the issues that arise when using AAs to implement these services and then present our experiences implementing one such service, namely *concast*, as an AA running on the ASP EE. The paper concludes with performance numbers obtained from an example audio-merge application that shows the viability of using AAs as specialized execution environments.

## 1 Introduction

Although the active network community has designed and largely embraced a common architectural framework for active networks [6], we are only now beginning to gain experience building active network applications that use the framework. This paper reports on our experiences building applications using the framework. In particular, we consider the issues involved with implementing higher-level services which we simply call *active services (AS)*. In some sense, active services are analogous to the upper layers of the IP protocol stack, offering more specialized services to applications than the basic network-level service (IP). We consider the requirements of active services, and discuss the suitability of the architectural framework for supporting these types of services.

The active network framework, illustrated in Figure 1, is quite general. It consists of a set of nodes (not all of which need be active) connected by a variety of network technologies. Each active node runs a *Node Operating System* and one or more *Execution Environments*. The *NodeOS* is responsible for allocating and scheduling the node's resources (link bandwidth, CPU cycles, and storage), while each *Execution Environment (EE)* implements a virtual machine that interprets active packets arriving at the node.

---

* Work sponsored by the Defense Advanced Research Projects Agency (DARPA) and Air Force Research Laboratory, Air Force Materiel Command, USAF, under agreement number F30602-99-1-0514, and, in part, by the National Science Foundation under Grants EIA-0101242 and ANI-0121438. The U.S. Government is authorized to reproduce and distribute reprints for Governmental purposes notwithstanding any copyright annotation thereon.

J. Sterbenz et al. (Eds.): IWAN 2002, LNCS 2546, pp. 191–203, 2002.
© Springer-Verlag Berlin Heidelberg 2002

Each EE defines a different virtual machine or "programming interface" on which *Active Applications (AAs)* can be built to provide a particular end-to-end service. One EE may provide a completely general virtual machine (e.g., Turing complete) while another provides a more restrictive virtual machine. However, in both cases the programming environment provided by an EE is intented to be useful to a wide range of applications. AA code, on the other hand, is designed to meet the specific needs of a particular application.

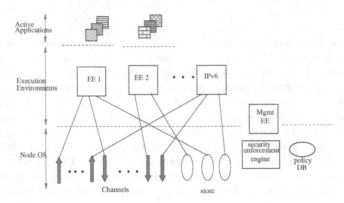

**Fig. 1.** The Architectural framework for an Active Network.

In the current Internet, applications are rarely built directly on the IP protocol. Instead, a *protocol stack* of higher-level protocols offer common/useful services (e.g., TCP, UDP, RPC, XDR, VPNs, RTP, etc.) needed by a wide range of applications. A similar need for high-level services arises in the context of active networks. These higher-level services may themselves be customizable, offering a special-purpose programming model and execution environment. For example, a "packet thinning" service (AS) might allow applications to define the function that selects packets to be dropped. Because packet thinning is a "higher-level" service, the packet thinning programming environment is much more restrictive (i.e., tailored to packet thinning) than the general programming environments offered EEs. Moreover, the active service typically performs the task of deploying and installing the user's function only at nodes that need it. Currently, the active network framework is agnostic about how such services should be implemented within the active network architecture.

One option is to implement each active service as an EE (see Figure 2(a)). However, given the specialized nature of an AS, it does not really fit the generic service model of an EE. Moreover, because EEs tend to be general-purpose programming models, development and deployment of an EE is considered a nontrivial, heavyweight exercise, and it is expected that the number of different EEs supported will be small. We expect there will be tens, possibly hundreds, of different active services which would imply a proliferation of EEs and clearly goes against the principles of the framework.

Another alternative is to implement each AS as an AA (Figure 2(b)). A clear advantage of this approach is that an AS does not need to re-implement resource allocation mechanisms or policies, but rather can leverage these services from the underlying EE. However, unlike normal AAs, "AAs used as ASes" must also support a programming environment capable of running application-specific code (i.e., other AAs). Followed recursively, this approach leads to AAs acting as execution environments for other AAs who may, in turn, act as even more specialized execution environments for other AAs (i.e., similar to a protocol stack).

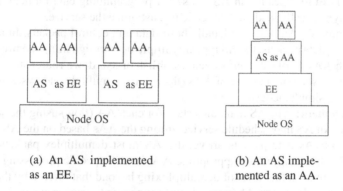

(a) An AS implemented as an EE.

(b) An AS implemented as an AA.

**Fig. 2.** AS implementation approaches.

A third approach which has not received much attention is to implement AS services as libraries linked against the AA. The resulting AA is then loaded into the EE and implements the AS functionality only for that application. This has some obvious drawbacks, including the need to load multiple copies of the same AS code, the AA being tied to an EE rather than an AS, and the difficulty of managing state and allocating resources across flows (AAs).

Given the drawbacks associated with implementing ASes as EEs or as a library, this paper explores the advantages and disadvantages of implementing ASs as AAs. We describe an implementation of a *Concast AS* executing as an AA on the ASP EE [1]. Although we show experimental results from an audio application that uses the concast AS, the main contributions of the paper is the experience gained implementing an AS as an AA. In particular, we found that many of the EE services had to be replicated by the AS. In some cases, re-implementing these services as an AA was challenging because the execution environment did not provide the necessary primitives to construct these services.

## 2   Active Services

The goal of an AS is to offer a high-level network service that is focused and limited in what it does (i.e., only supports a limited set of methods), but yet can be customized to the particular needs of the application. Example AS services include reliable multicast services, intelligent discard (congestion) services, mobility services, interest filtering, intrusion detection, or network management queries. In this paper, we consider a *concast* active service. Concast is roughly the inverse of multicast, merging packets as they travel from a set of senders to a receiver. Although we focus on the concast service, many of the issues that we address are typical of other ASes including:

**Restricted Programming Model:** An AS must "sandbox" the AA code thereby limiting it to do only what is allowed by that specific service.

**AA Code Loading:** Because an AS is itself a programming environment, it must be able to dynamically load an AA's code to customize the service.

**Data and Control:** An AS must handle both data and control packets. In addition to processing data packets, the AS typically implements a signalling (control) protocol by which AAs invoke the service and establish or tear down router state.

**Simultaneous Use:** Any number of AAs (flows), from multiple end-systems, may use the service simultaneously.

**State Management:** The AS maintains state for each AA (flow) using the service.

**Scheduling:** An AS may schedule service among the AAs based on the AS's policy.

**Packet Demux:** As data packets arrive, the AS must demultiplex packets (possibly after AS processing) to the appropriate AA for AA-specific processing. Note that this represents another level of demultiplexing beyond that offered by the EE.

**Interrupt Processing:** The AS may schedule timeouts to trigger events in the future (e.g., the transmission of packets at fixed time intervals).

**Shared State:** Some ASes may allow different AAs to communicate or share state.

Because these characteristics are similar to those of an EE, it is tempting to say an AS should just be an EE. However, there are subtle, but important, differences between ASes and EEs (in addition to the arguments given earlier against implementing an AS as an EE). First, an AS programming model is more focused and restrictive than an EE programming model. Second, an AS is not interested in re-implementing basic EE functionality like routing and forwarding. Third, an AS does not want to implement resource policies and mechanisms – things a NodeOS or EE must do. In short, an AS is not an EE. Instead, an AS should build on, and leverage, EE services.

## 3  The ASP EE

To explore the issues involved with implementing higher-level services (i.e., ASes), we implemented a *concast AS* as an AA running on the ASP EE. The ASP EE offers a Java-based programming environment that is designed to assist in the development and deployment of complex control-plane functionality such as signalling and network management. ASP implements the "strong EE model" [1] which means it essentially offers a user-level operating system to AAs. Although the underlying NodeOS and Java programming language provide operating system services, ASP tailors these services to its own goals and objectives. The primary enhancements are to the network I/O constructs. ASP uses *Netiod* to replace the socket interface with the active network input and output channel abstractions [6]. These channels effectively filter incoming and outgoing packets into streams. The interface between AAs and the ASP EE is called the *protocol programming interface (PPI)*, and allows AAs to access the active node's resources, such as storage, CPU cycles, and bandwidth, through the ASP EE. ASP also defines an interface to user applications (UA) on end systems. The UA can be thought of as 'adjacent' to the ASP EE and communicates with ASP using a TCP connection.

Being Java-based, ASP provides a certain amount of security that is inherited from the scope rules of the Java language. The ASP EE increases this protection by employing a customized Java Security Manager, which protects the EE and the JDK from aberrant or malfunctioning AAs. However, because each AA is assumed to be written by a single AA designer, there is no mechanism to protect certain parts of the AA from other parts of the AA (which we will see is needed by an AS).

ASP does not currently support in-band code loading, but rather supports an out-of-band loading mechanism. The ASP EE requires that active packets include an *AAspec*, which supplies a reference to the AA code to be run.

The ASP implementation executes each AA in a distinct process. Each process may contain multiple Java threads defined using the *AspThread Class*, which is akin to the Java thread class with some slight enhancements.

The ASP EE recognizes two methods for routing active packets: native (IP) connectivity and virtual (VNET) connectivity. Native IP connectivity allows AAs to use the IP protocol stack for routing. The ASP EE provides virtual connectivity via virtual links constructed using UDP/IP tunneling. The ASP EE's ability to provide routing functionality relieves the supported AAs from the responsibility of directly determining, maintaining, and using routing tables and protocols.

# 4   An Example Active Service: Concast

We begin with a brief overview of the *concast service*, and then describe our experiences implementing the service as an AA running on ASP.

## 4.1   The Concast Service Abstraction

*Concast* is roughly the inverse of IP multicast. Instead of a sender multicasting packets to a group of receivers, a *group of senders* send packets toward a single receiver. The concast service merges these packets together using a user-specified merge function [3].

A *concast flow* is uniquely identified by the pair $(G, R)$ where $R$ is the receiver's (unicast) IP address and $G$ is the *concast group identifier*. $G$ represents the set of senders that wish to communicate with $R$. Concast packets are ordinary IP datagrams with $R$ in the destination field and $G$ in the IP options field (in a *Concast ID* option). The packets delivered to $R$ are a function of the packets sent by the members of $G$. Concast-capable routers intercept and divert for processing all packets that use the Concast ID option.

The concast abstraction allows applications to customize the mapping from sent messages to delivered message(s). This mapping is called the *merge specification*; it controls (1) the relationship between the payloads of sent and received datagrams, (2) the conditions of message forwarding, and (3) packet identification (i.e. which packets are merged together). The merge specification is defined in terms of four methods **get-Tag**, **merge**, **done**, and **buildMsg**. The concast framework allows users to supply the definitions of these "merge-spec" functions using a mobile-code-language (currently Java). Once defined, a merge specification is injected into the network by the receiver and is pulled toward the senders as they join the group (using the *Concast Signalling Protocol (CSP)*). Additional details about the CSP protocol and merge specification can be found in [4, 5, 13].

Incoming packets are classified into flows based on $(G, R)$. The flow's **getTag** function is applied to the packet to obtain a tag that identifies the equivalence class of packets to which the packet belongs. The **merge** function is then invoked to merge the current packet with the previous packets. The **done** predicate then determines whether the merge operation is complete. If so, **buildMsg** is invoked to construct an outgoing message from the merged state that is forwarded toward the receiver.

The two primary components of the concast service are the concast signalling protocol (CSP) and the merge processing framework. In our native (Linux) kernel implementation [5], we implemented these two components as UNIX processes: one process acted as the *Concast Signalling Protocol Daemon (CSPD)* and any number of processes executed *Merge Daemon (MergeD)* code (depending on the number of active concast flows). The CSPD handled all control messages (e.g., join and leave requests) and also "forked and exec'd" a MergeD process each time a new flow was initiated. Because each MergeD was a distinct process, MergeDs could not intentionally or accidentally affect one another.

### 4.2   Implementing Concast as an AS

As noted above, the concast service exhibits many of the characteristics typical of an active service(see Section 2). It needs to support multiple concast flows (i.e., must accept and process packets from multiple concast flows simultaneously). It must dynamically load code for each concast flow (i.e., the merge specification). It must handle both data traffic (packets to be merged) and control traffic (concast signalling messages to join/leave groups, pass merge specifications, or refresh state). For data traffic, it must ensure packets are delivered to, and processed by, the appropriate merging code. It also needs to focus (restrict) the programming model (i.e., only allow definition of the four merge-spec functions). Although it does not implement a scheduling policy for the processor, it does restrict the state that can be accessed and manipulated by the application code (i.e., merge spec). Finally, it needs to support interrupt processing for timer-triggered events (e.g., applying periodic processing or forwarding packets after fixed time intervals).

Figure 3 illustrates our design for the concast AS, and shows how it fits into the overall active network framework.

The first challenge arises from the need to support multiple concast flows (groups) using the service simultaneously (i.e., the AS must support multiple MergeDs executing simultaneously). Unfortunately, ASP provides no mechanism by which an AA can "fork and exec" a new AA to perform parallel MergeD processing.

To achieve the desired parallelism, we implement the CSPD and all MergeDs as distinct *ASP threads* inside a single AA. Assigning each merge specification to a different ASP thread allows us to maintain clean separation of the (independent) MergeDs from one another. Each thread maintains merge state specific to that flow. It also offers the

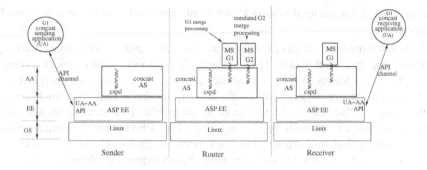

**Fig. 3.** The Concast Service implemented as an AA on ASP.

opportunity to execute the application's customization code (i.e. merge spec) in parallel. Although parallel execution is not necessarily important for concast merge specs, it is important for other active services that do not always process packets to completion. The alternative, using a single thread to execute the customized code from all applications, is possible, but complicates the code and can produce unnecessary blocking, particularly when a packet cannot be processed to completion.

Given multiple merge specifications executing simultaneously in distinct ASP threads, a related issue is the problem of getting incoming concast packets to the appropriate MergeD (i.e., ASP thread). As noted earlier, ASP allows an AA to establish one or more in-channels (classifiers) that define the set of packets the AA should receive. Because ASP wants to deliver packets to the AA in the order that the packets were received (across all in-channels), ASP delivers all packets to the main ASP thread (regardless of the in-channel that the packet is associated with). In other words, ASP does not record which thread opened the in-channel and thus cannot deliver packets for a channel directly to the thread that requested the packets.

Because ASP delivers all packets directly to the main thread, we had to implement the ability to demultiplex packets to the appropriate threads (i.e., MergeDs). To do this, we used the main ASP thread as a "packet dispatcher", examining each packet's receiver address R and group id G, and demultiplexing the packet to the appropriate MergeD thread. If the packet is a CSP (control) packet rather than a data packet, the main thread immediately applies the CSPD processing, possibly creating or terminating a thread (i.e., a MergeD) in response to a CSP join or leave message. Ideally, the underlying EE would have offered the ability to assign in-channels to threads rather than AAs (thereby avoiding this demultiplexing overhead).

This raises the next issue: interthread communication. Interthread communication was implemented using shared java monitor queue objects between the dispatcher thread and the MergeD threads. By wrapping the queue in a java monitor object, we were able to ensure mutually exclusive access to the queue. The dispatcher thread read the packet, identified the appropriate queue object, and then used the monitor to lock and append the packet to the queue. It then signalled to inform the MergeD thread that a packet was available. In this case, the base language offered the support we needed. However, if the base language did not offer this support, we would have had to implement our own IPC mechanism.

Note that our design suffers from the same protection and security issues that plague EE's that support multithreading [14]. Systems that support threads typically allow all threads to access the same data (i.e., shared access). As a result, the merge-spec for one flow could conceivably interfere with, kill off, or otherwise disrupt the MergeD of another flow. ASP actually protects AAs from one another by executing each in its own JVM (i.e., its own virtual memory process). The ANTS EE [14] faces a similar problem, namely ensuring that data deposited by packets of one protocol cannot be accessed by packets in another protocol. In Java, accessing objects requires a reference to the object, and such references can only be obtained through legitimate channels (i.e. it is not supposed to be possible to convert arbitrary data to references). ANTS arranges things so that only authorized code can obtain a reference to an object, because the reference is tied to a cryptographic hash of the code itself! Thus, in ANTS, objects are secure from tampering by other flows. In the case of our concast AS, the MergeD code executes in the context of an ASP thread that does not know about, interact with, or exchange references with any other threads; consequently, data and packets referenced by one thread are not accessible to another thread, thereby guaranteeing protection between threads.

One example where the ASP EE restricts the base-language primitive in a way that hinders the development of an AS is the ASP class loader. Rather than reinvent our own class loader, we wanted to utilize the ASP class loader. Although the ASP class loader greatly simplified our implementation, it also limits our implementation in some minor, but important, ways. In particular, to prevent an AA from dynamically loading code from anywhere, the ASP class loader requires that all loadable code (including the AA code) reside in one of a set of pre-specified directories that are associated with the AA. However, in the case of an AS, the code will reside at a location of the application's choosing (i.e., an arbitrary location). Because ASP restricts the location where code can be loaded from, its loader is not well suited for our application. To get around this problem, we enabled applications to deposit their code at the URL for the concast AS. This is clearly not an elegant or secure solution.

A related problem is that of restricting the programming environment offered by an AS. ASP uses a customized Java Security Manager to protect ASP from the AAs. However, the goal is security/safety rather than restricting the programming model – almost all Java functionality is available to the AA. Although we have not implemented one, our concast AS could also develop its own specialized security manager to disable certain functionality. Although this would limit the programming model, it may not be powerful enough to define the specialized programming models offered by certain ASes.

One of the biggest problems we faced was implementing timeouts. The concast programming environment supports the notion of timeouts to trigger periodic processing of data or transmissions of already merged data. This feature is used, for example, in the audio merge example described in Section 5. The primary problem we encountered was the inaccuracy of the timeout mechanism. Although we still have not completely identified all the contributors to the problem, one of the major factors was the layering that caused the timing errors to accumulate. In concast, timeouts are implemented using an ordered per-merge-spec timeout list. Each time the merge framework processes a message it checks for any pending timeouts, processes those, and then issues a timed wait for the next packet (based on the next timeout to occur). Like most timeout mechanisms, this implementation does not guarantee that the timeout processing will occur precisely at the point of the timeout, but rather "soon after" the timeout. The concast timeout mechanism itself relies on the underlying timeout mechanism provided by Java (i.e., a timed wait). The Java mechanism itself relies on the timing mechanisms of the underlying operating systems. Because each layer depends on the "imprecise" timeout mechanism of the layer below it, the accumulation of errors often lead to unacceptable accuracy at the AS level.

We experienced variance as high as 10ms which was beyond the tolerance of our audio example. Even if a single error of a few milliseconds could be tolerated, repeated inaccuracy over time led to an accumulation of errors that were unacceptable. Although our measurements show that a significant portion of the inaccuracy can be attributed to the timed waits in Java, it appears that some of the inaccuracy accrued from thread switching and packet demux. If the AS can be implemented with a single thread, we expect the accuracy would improve, but will still suffer the effects of layering.

In short, some aspects of ASP made it easy to offer the necessary services and features, while other aspects were not well-suited as ASP is currently designed. However, with feedback about the characteristics needed by the average AS, EE's could be designed to offer the appropriate support so that implementing an AS as an AA is both viable and attractive.

# 5   Experimental Results

To evaluate the performance aspect of an AS implemented as an AA, we developed an audio merge application that employs our concast AS service to do the merging. In what follows, we describe the audio merging application, our experimental environment and setup, and results obtained from a running system.

## 5.1   An Audio Merge Example

To test our concast AS and measure its performance characteristics, we constructed an audio application that merges together multiple audio flows to form a single "combined" audio flow that is delivered to a (concast) receiver. This type of audio merging/mixing might be used in conference calls or teleconferencing, distributed music performances, or it might be used to merge audio from distributed sensors used for audio surveillance.

The basic idea behind our approach is that each sender transmits a u-law encoded audio stream at a fixed rate (i.e., 64 kbps in our case) to a receiver (i.e., the merge destination). The goal is to combine the audio streams and play them as a single audio stream at the receiver. The receiver creates the concast flow by installing an "audio merge specification". The audio merge specification executes at each router combining an audio packet from each of the incoming flows to create a single outgoing packet. In our case, merging is achieved via a simple function that first maps u-law encoded audio samples to a-law encoded samples. It then combines the samples by adding a-law values from different flows together and truncating the results if the result exceeds the maximum allowable value. Finally, it maps the resulting a-law sample back to a u-law encoded sample. The merge function also includes timeout processing. The intent is that as packets arrive, the merge function merges the incoming packet into the state accumulated thus far. At regularly scheduled 125 ms intervals, the node packages the accumulated state and forwards it toward the next downstream router. Senders transmit a 64 Kbps audio stream, sending a packet every 125 ms where each packet contains 1000 eight-bit samples. The intermediate nodes are impervious to packet loss; they blindly forward whatever they accumulate during the 125 ms time period. If no packets were received, the node forwards nothing during that interval.

Because of the difficulty realizing precise timeouts in the AS, we implemented a version of the MergeD that buffered up to 3 packets from each upstream neighbor. Every 125 ms the MergeD removed the first buffered packet from the queue for each upstream neighbor, merged them, and forwarded the result. Normally, if a node received two consecutive packets from the same sender during a single 125 ms interval, it would discard the second packet (because we do not want to merge together packets from the same sender). By employing a small amount of buffering at each node, the merge function became much less susceptible to packet drops resulting from timing errors.

## 5.2   Experimental Setup

We tested our concast AS and audio merge applications using the topology shown in Figure 4. We implemented the running system and topology using the EMULAB system at the University of Utah [8]. We used pre-recorded audio files as the audio input to the senders, and the receiver wrote an audio output file.[1] The last-hop link in the topology

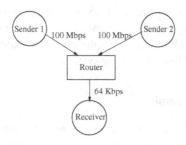

**Fig. 4.** Topology of our test network.

was intentionally designed to be a bottleneck link; the 64 Kbps link can support at most one audio flow at a time. Our experiments consisted of two senders, one router and one receiver.[2] Each node ran the Linux operating system, ASP, and our concast AS.

---

[1] We also tested a live playout.

[2] We have tested with more elaborate topologies and more senders, but two senders is sufficient to illustrate the AS performance.

Senders and receivers use the *User Application (UA)* and the *UA-AA* API of ASP to send active packets to and receive active packets from the ASP AA. ASP nodes communicated using ASP's *VNET* network interface. A concast flow is created when a concast receiving application sends a CREATE request to a local concast AS using the UA-AA API. The CREATE request specifies the concast address and merge specification and indicates the receiver's willingness to accept data sent to this concast address. Subsequently, the concast sending applications can join this concast flow by sending a JOIN request to their local concast AA (using the UA-AA API). This triggers the concast signalling protocol which "pulls down" the merge spec and loads it via the ASP class loader.

As a point of comparison, we also implemented a unicast version of the merge processing. In the unicast case, the senders transmitted (independent) unicast flows to the receiver; the receiver then merged the packets together before playout. Because the unicast flows were not merged until they reached the receiver, their combined load exceeded the bandwidth of the last-hop link.

## 5.3  Performance Results

Figure 5a compares the number of packet drops that occur under the the the concast implementation (with a buffer size of 3 packets per sender) and the unicast implementation. The more packets dropped, the worse the sound quality. One sender transmitted melodic background music while the other sender transmitted someone talking. In the unicast implementation, the music was choppy and the speaker could not be understood. The concast implementation with buffering completely eliminates all packet drops, particularly the drops caused by the timing inaccuracies of the AS. The resulting sound quality was very good, because there are almost no packet drops.

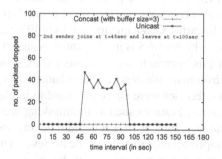 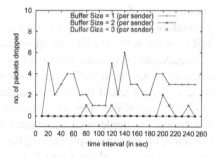

**Fig. 5.** (a) Packet drops for Concast vs. Unicast. (b) Packet drops for Concast with various buffer sizes.

Figure 5b illustrates the effect buffering has on the performance of the concast AS. Because timing inaccuracies can lead to "bursts" of up to 3 packets (i.e., from the same sender arriving in the same 125 ms interval), buffer sizes of only 1 or 2 packets will drop some packets. These packet losses produced a periodic "ticking" sound in the audio. A buffer size of 3 packets per sender eliminated all packet drops in our experiment and produced good sound quality.

# 6  Related Work

Past work in active nets [2] has focused largely on the NodeOS and EE levels. EEs such as ANTS [14] and ASP [1] are Java-based environments each leveraging the Java language in slightly different ways. Had we selected ANTS as the development environment as opposed to ASP, most of our approach would have remained the same given the fact that they are both Java-based. Other EEs, such as PLAN [9, 12], could also have been used. Although we have relatively little hands-on experience with the PLAN EE, we expect that it would have been more helpful with defining the restricted programming model desired by the AS. The CANEs [7] EE offers a more restricted programming model and may have presented some special challenges to implementing an AS as an AA.

Relatively few active applications have been implemented. The few that have been implemented focus on a total solution to a problem such as video processing [11], distributed interest filtering [15], and network management [10]. Each of these executes as a complete AA. Concast is one of the first higher-level AAs that actually offers a service and programming environment to other applications.

# 7  Conclusions

In this paper we addressed the issue of implementing higher-level active network services; services that are more specialized than EEs, but yet are programmable like EEs. We call these middleware services *active services* (AS). We outlined three methods for implementing active services: EEs, AAs, and libraries. Given the drawbacks of EEs and libraries, we decide to explore the approach in which an AS is implemented as an AA.

To understand the issues associated with this approach, we implemented the *concast* AS (as an AA) in the ASP execution environment. We described the characteristics of an active service, and showed which characteristics were straightforward to implement in ASP and which ones presented problems. Features such as multitasking were straightforward to implement, while other features, such as packet demultiplexing and code loading required re-implementing (or adapting) these services at the AS level. The characteristic that posed the biggest challenge was timer management. We presented experimental results from an audio application that show the performance of an AS can be acceptable for realtime packet processing.

In summary, we've pointed to EE features that were well suited for supporting active services, and other features that could be designed better. Our experience indicates that implementing active services as AAs is a viable approach, both from an ease-of-implementation standpoint and a performance standpoint. As we gain more experience with active services, time will tell if AAs are the correct approach for implementing active services.

# References

[1] Robert Braden, Bob Lindell, Steven Berson, and Ted Faber. The ASP EE: An Active Network Execution Environment. In *DARPA Active Network Conference and Exposition (DANCE) 2002*, May 2002.

[2] K. Calvert, S. Bhattacharjee, E. Zegura, and J. Sterbenz. Directions in active networks. *IEEE Communications Magazine*, 36(10):72–78, October 1998.

[3] K. Calvert, J. Griffioen, B. Mullins, A. Sehgal, and S. Wen. Concast: Design and implementation of an active network service. *IEEE Journal on Selected Areas of Communications*, 19(3):426–437, March 2001.

[4] K. Calvert, J. Griffioen, A. Sehgal, and S. Wen. Concast: Design and implementation of a new network service. In *Proceedings of 1999 International Conference on Network Protocols, Toronto, Ontario*, November 1999.

[5] K. Calvert, J. Griffioen, A. Sehgal, and S. Wen. Implementing a concast service. In *Proceedings of the 37th Annual Allerton Conference on Communication, Control, and Computing*, September 1999.

[6] Kenneth L. Calvert. An Architectural Framework for Active Networks, 2001. DARPA Active Nets Document, http://protocols.netlab.uky.edu/ calvert.

[7] Kenneth L. Calvert and Ellen W. Zegura. Composable Active Network Elements Project. http://www.cc.gatech.edu/projects/canes/.

[8] Flux Group. Emulab Network Testbed. Computer Systems Lab, University of Utah, http://www.emulab.net/.

[9] Michael Hicks, Pankaj Kakkar, T. Moore, Carl A. Gunter, and Scott Nettles. PLAN: A Packet Language for Active Networks. In *Proceedings of the International Conference on Functional Programming*, 1998.

[10] A. Jackson, J. Sterbenz, M. Condell, and R. Hain. Active Network Monitoring and Control: The SENCOMM Architecture and Implementation. In *DARPA Active Networks Conference and Exposition*, pages 379–393, San Francisco, May 2002.

[11] R. Keller, S. Choi, M. Dasen, D. Decasper, G. Fankhauser, and B. Plattner. An Active Router Architecture for Multicast Video Distribution. In *IEEE INFOCOM*, Tel-Aviv, Israel, March 2000.

[12] Jonathan T. Moore, Michael Hicks, and Scott Nettles. Practical Programmable Packets. In *IEEE INFOCOM*, Anchorage, AK, April 2001.

[13] Amit Seghal, Kenneth L. Calvert, and James Griffioen. A Flexible Concast-based Grouping Service. In *Proceedings of the International Working Conference on Active Networks (IWAN) 2002*, December 2002.

[14] David J. Wetherall, John V. Guttag, and David L. Tennenhouse. ANTS: A Toolkit for Building and Dynamically Deploying Network Protocols, 1998.

[15] S. Zabele, M. Dorsch, Z. Ge, P. Ji, M. Keaton, J. Kurose, and D. Towsley. SANDS: Specialized Active Networking for Distributed Simulation. In *DARPA Active Networks Conference and Exposition*, pages 356–365, San Francisco, May 2002.

# ROSA: Realistic Open Security Architecture for Active Networks

Marcelo Bagnulo, Bernardo Alarcos, María Calderón, Marifeli Sedano

Departamento de Ingeniería Telemática, Universidad Carlos III de Madrid
Av. Universidad 30, 28911 LEGANES, MADRID
{marcelo,maria}@it.uc3m.es
Área de Ingeniería Telemática, Universidad de Alcalá
28871 Alcalá de Henares, MADRID
{bernardo,marifeli}@aut.uah.es

**Abstract.** Active network technology enables fast deployment of new network services tailored to the specific needs of end users, among other features. Nevertheless, security is still a main concern when considering the industrial adoption of this technology. In this article we describe an open security architecture for active network platforms that follow the discrete approach. The proposed solution provides all the required security features and it also grants proper scalability of the overall system, by using a distributed key generation algorithm. The performance of the proposal is validated with experimental data obtained from a prototype implementation of the solution.

## 1 Introduction

Active networking technology[1] has already proven to be a powerful approach when fast deployment of new protocols and services is needed. However, security risks introduced by its own nature are a major concern when evaluating the usage of this technology in public environments. Furthermore, heavy security measures can preclude deployment in real scenarios because of the imposed overhead in terms of processing, bandwidth and/or latency. So, in order to achieve a deployable active network architecture, the security solution must not only provide protection from all the detected threats but it must also grant the scalability of the system. In this article, we will present ROSA, a Realistic Open Security Architecture for active network platforms that follow the discrete approach [2], which can fulfill both requirements thanks to a distributed key generation algorithm and to architectural features of the discrete approach platforms.

The remainder of this article is structured as follows. In section 2, an introduction to discrete approach to active networks is presented. In section 3, the security solution requirements are detailed, including threats assessment and scalability requirements. Next, section 4 provides an overall description of the proposed security architecture. In section 5, implementation is described and performance results are discussed. Section 6 is dedicated to related works. Finally, section 7 is devoted to conclusions.

J. Sterbenz et al. (Eds.): IWAN 2002, LNCS 2546, pp. 204–215, 2002.
© Springer-Verlag Berlin Heidelberg 2002

## Active Networks The Discrete Approach

There are two different approaches to provide dynamic network programmability to active networks. Some active network platforms follow a discrete approach. This means packets do not include the code to be executed in the Active Routers but a separate mechanism exists for injecting programs into an Active Router, such as a *Code Server*. Other active network platforms follow an integrated approach and packets, called capsules, include not only user data but the code used for the forwarding of the packet as well. We will next present three active networks that follow the discrete approach DAN, SARA and ASP. These platforms are compatible with ROSA. Finally, we will describe the active packet exchange.

### Discrete Approach Platforms

DAN[ ], which stands for Distributed Code Caching for active networks, has been developed by the Washington University of St Louis and by the Computer Engineering and Network Laboratory of Zurich. DAN is an Execution Environment EE that is running in the high performance Active Network Node ANN[ ]. The proposed framework mainly includes the following components, an Active Module Loader which loads the active modules authenticated and digitally signed by their developers from well known code servers using a lightweight network protocol, a Policy Controller which maintains a table of policy rules set up by an administrator, a Security Gateway which allows denies active modules based on their origin and developer by analyzing their digital signatures authentication information, a Function Dispatcher which identifies references to active modules in data packets and passes these packets to their corresponding function implementations, and a Resource Controller for fair CPU time sharing among active functions.

The next active network platform considered is SARA, Simple Active Router Assistant[ ], which is an active router prototype developed by the University Carlos III de Madrid in the context of the IST project GCAP[ ]. It is based on the router assistant paradigm, meaning that active code does not run directly on the router processor but on a different device called assistant, which is directly attached to the router through a high speed LAN. Hence, the router only has to identify and divert active packets to its assistant. Active packets are identified by the router alert option, enabling active router location transparency, since active packets need not be addressed to the active router in order to be processed by it. After requested processing is performed by the assistant the packets are returned to the router in order to be forwarded. The active code needed to process active packets is dynamically downloaded from Code Servers when it is not locally available in the assistant. In this way safety can be checked in advance since only registered code proved harmless is stored in code servers. SARA is available in two platforms. One fully based on linux [ ] and a hybrid platform where the router used is an Ericsson Telebit AXI running a kernel adapted to work with an active assistant.

Finally we will consider one of the principal EEs running within ABone[ ], which is ASP[ ] from the University of the Southern California Information Sciences Institute. In the implementation proposed in this EE the active code is downloaded from a set of known secure code servers. An important contribution of ASP is the support of persistent active applications that may have long lived execution threads.

### Discrete Platform Packet Exchange

In order to present the security architecture we will first introduce the packet exchange performed so we can detect the requirements imposed by security and scalability concerns.
The elements involved in the packet exchange are:
*Source.* User terminal that generates traffic and uses the active services.
*Destination.* It is the terminal that *Source* addresses its traffic to.
*Active Router.* It is a router capable of processing active packets. It is also able to obtain the active code needed.
*Code Server.* It is the active code repository that serves the *Active Routers.*

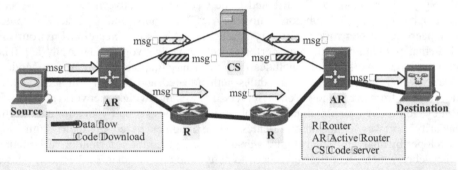

**Figure** Services Network Architecture

The packet exchange description depicted in figure is described next. When *Source* needs special active processing for a flow of packets between itself and *Destination*, it must send packets (msg) addressed to *Destination* containing the identification of the active code that it desires to be executed. When a packet reaches the first *Active Router*, it is inspected and the identification of the active code is extracted. If the active code is locally available at the *Active Router* it performs the requested process and then forwards the packet (msg). If the needed active code is not locally available the *Active Router* requests it from the *Code Server* (msg). The *Code Server* then sends the requested code to the *Active Router* (msg) which now processes the packet and forwards it to the next hop. The same procedure is executed by all the *Active Routers* along the path (msg, msg) until the packet reaches *Destination*, where the packet is received (msg). Next active packets of this flow will presumably follow the same path, so the *Active Routers* will be capable of processing them without needing to request the code from *Code Servers* again.

## Security Architecture Requirements

In this section we will present the different requirements imposed on the security architecture We will first start by stating the security requirements and then we will describe other general requirements specially emphasizing the scalability aspects

### Security Requirements

Security requirements imposed by active networks have already been detailed in several documents [   ] So we will not perform an exhaustive analysis here but we will only present the final security requirements from each element's perspective

From the *Active Router's* perspective authorization is a key requirement It is relevant that the active code loaded into the routers is provided by an authorized *Code Server* and not from an unauthorized source In addition the code integrity must be preserved while it is transmitted from the *Code Server* to the *Active Router* In addition the *Active Router* must be able to verify that the user that is requesting the code *i.e. Source* is authorized to execute it at this moment

From the *Code Server's* perspective it must be able to authenticate *Active Routers* that are requesting active code since not all the code will be available to all routers Furthermore the security solution must provide confidential code transfer in order to prevent unauthorized parties from inspecting the delivered code

From the *Source's* perspective it must be able to be certain no other user is requesting active services on its behalf It must also be the only one capable of controlling its active services meaning that no other user is capable of introducing new active packets or modifying active packets sent by *Source* interfering with the requested active service In addition to authentication features it is also important to provide non repudiation this is specially important when active services will be provided in a commercial fashion

From the *Destination's* perspective there are no requirements since it does not demand active services from the network It should be noted that end to end security is out of the scope of this security solution

### Other General Requirements

Besides security requirements the security architecture must also meet scalability and performance requirements which are reflected next

- Zero user knowledge at the *Active Routers* In order to build a manageable solution user management must not be performed on each and every *Active Router* A database containing all the users information including access rights would be the preferred solution
- Path transparency It must not be required that the *Source* be aware of which *Active Routers* are in the path used to transport packet towards the *Destination* In addition it must not be required that each node in the path has knowledge of its active neighbors These requirements are needed to grant the scalability

performance and flexibility of the active network since hop by hop authentication is considered to be incapable of providing the mentioned features

# ROSA Security Architecture

In this section we will present the proposed security architecture We will first consider *Source* authorization issues evaluating the different authorization paradigms available and then inferring which one is the most appropriate for this particular problem Then we will consider the code downloading security and non repudiation issues Finally we will present the overall solution step by step

## *Source* Authorization

### Authorization Paradigm

A key feature that must be provided by the security architecture is authorization i e *Sources* must be authorized to execute the solicited code on *Active Routers* There are two authorization paradigms that can be used authorization based on access control lists or authorization based on credentials The first paradigm is based on the existence of an access control list locally available or in a remote location that must be queried every time an *Active Router* receives an active packet sent by *Source* in order to validate the *Source's* permissions In this case the identity of the requesting party must be authenticated in order to prevent impersonation This approach then requires that the requested device *Active Router* has information about *Sources* and permissions or it imposes a communication with an authorization server every time a *Source* sends an active packet The second paradigm demands that every time *Source* sends a packet a credential that proves the *Source's* permissions must be presented Then the requested device *Active Router* only needs to verify the credential However credential generation and distribution may be more than a trivial task

The solution proposed in this paper will be designed based on the second paradigm since we consider that it provides better scalability attributes

### Considering the Usage of Public Key Cryptography

In order to allow the intended use a credential must contain verifiable authorization information i e the permissions granted to the holder of the credential In addition it must be possible to verify that the issuer of the credential is a Valid Issuer i e that it has the authority to grant these permissions It is also critical to validate that the user that is presenting the credential is the same user that the credential was granted to

In order to fulfill the above stated characteristics of a credential public key encryption can be used So a credential containing the *Source's* permission and the *Source's* public key is signed by the Valid Issuer Then the *Active Router* must be capable of verifying the authenticity of the credential and also it must be capable of verifying that the requesting user has the private key that corresponds to the public key included in the credential. This mechanism provides all the required features, but the usage of public key cryptography is very demanding in terms of processing.

## Authorization and Key Generation Solution

In order to obtain a less demanding solution, symmetric key cryptography can be used. However, building a similar system using symmetric key would require the usage of two different symmetric keys: a first one shared by the Valid Issuer and the *Active Routers* and another key shared by *Source* and the *Active Routers*. This system would still demand two cryptographic verifications and it would present the additional problem of key distribution. So, in order to improve the scalability of the solution, we will next explore the possibility of using only one symmetric key, shared by the Valid Issuer, *Source* and the *Active Routers*.

The requirements imposed on this key are:

– Different keys for different *Sources*, i.e. the key must be linked to a *Source*.
– Different keys for the same *Source* at different moments, i.e. key validity period.
– Different keys for different active codes requested by the same *Source*, i.e. the key must be linked to an active code/active service.

Therefore, the key K issued by the Valid Issuer is linked to a *Source*, an active code and a validity period.

Then, if K is used for generating an HMAC [ ] included in active packets that requests the execution of a particular active code, the active packets themselves play the role of credentials. Basically, an *Active Router* receives an active packet that includes the requested code identification, the *Source* identity, the time when the active service was requested, the requested period and an HMAC. Then, if the *Active Router* has a valid key K linked to the *Source*, the requested code identification and the validity period, it can verify the authenticity of the active packet without any further information. This mechanism imposes the usage of an *Authorization Server* (the Valid Issuer role) that generates the keys K. So, in order to execute a code in the network, *Source* must obtain the correspondent key K from the *Authorization Server* in a secure way. This is not a time critical task, since it is only performed when the service is requested and it is possible to be executed in advance. However, once the service is authorized and the key K is generated, the *Authorization Server* must communicate it to all *Active Routers* in the network, so they are aware of the new authorization. This is not the most scalable solution, because of the amount of communications needed between the *Authorization Server* and the *Active Routers*.

We will next present an improved solution that minimizes the required interactions. The basic idea is that the key K can be almost autonomously generated in every *Active Router* when it is needed. In order to achieve this, we will associate a key $Kc_i$ to every active code $C_i$ that can be loaded in the *Active Routers*. These keys $Kc_i$ are known by the *Code Server* and by the *Authorization Server*. Then, when a *Source* S requests authorization for the execution of code $C_i$ at a moment T and for a period P, the *Authorization Server* generates the secret key K as the HMAC of the concatenation of the parameters $Kc_i$, S, T, P and $C_i$:

$$K = HMAC (Kc_i, S, T, P, C_i)$$

The key K is then transmitted to the *Source* so it can generate the HMAC that will be included in active packets with it If we analyze the characteristics of K we can see that K is linked to an active code Ci Kci K is linked to a *Source* S K has a validity period T T P K can not be generated by the *Source* since it does not have Kci In addition the *Code Server* can attach Kci to the active code when this is confidentially downloaded to the *Active Routers* So the *Active Routers* are capable of regenerating K without contacting the *Authorization Server* every time an active packet arrives or when a new *Source* requests an already downloaded code The *Active Routers* have all the information needed to generate K ie S T P and Ci are included in all active packets and Kci is obtained when they download the code from de *Code Server*

Note that since the solution is based on shared secret keys the security level of the solution can be defined by setting the number of parties that share the Kci keys authorized *Active Routers* for a given code Ci and the frequency with which Kci keys are changed A re keying procedure based on the usage of multiple overlapped keys has been defined which can be easily tuned to obtain the requested security level It should also be noted that since Kci are stored in routers it is assumed that routers have some form of secure storage capabilities

## Code Downloading

Another key feature that must also be provided is a secure way to download code and Kci keys from the *Code Server* into the *Active Routers* However this is not as time critical as user authorization since it is only performed once when the first packet arrives to an Active Router The subsequent packets will benefit from a cached copy of the code and the Kci So a protocol that allows a secure communication between two parties is needed We will use TLS [ ] since it provides all the needed features Then both *Code Server* and *Active router* must have a digital certificate public key cryptography is used and a TLS session is established between the *Code Server* and the *Active Router* before the code is downloaded

## Non repudiation

When the user is requesting a service commercial and legal issues may be involved so non repudiation is relevant Furthermore the security architecture can be used to enable charging mechanisms In this case non repudiation is considered as an important asset In order to assure non repudiation public key cryptography must be used when the user requests authorization to the *Authorization Server* as will be described in the following section

## The Security Solution Step by Step

In this section we will describe the complete mechanism illustrated in figure First step in figure the *Source* requests authorization to the *Authorization Server* to execute an active code Ci in the network getting an active service This request is done in a secure way meaning that public key cryptography and digital certificates are used by both parties Therefore *Source*'s request is signed with the private key of *Source* and its digital certificate is also included This request is encrypted with the public key of the *Authorization Server* Then the *Authorization Server* after receiving and verifying the request it generates K as the HMAC of the *Source*'s identification S the requested code's identification Ci the key associated to this code Kci the service request time T and the validity period requested by *Source* P as we presented in section Then the *Authorization Server* sends a signed message containing K The message is encrypted with the public key of *Source*

The *Source* decrypts the message and obtains K Then step in figure it generates active packets that include its own identification S the service request time T the validity period P and an identifier of the requested active code Ci This message includes an HMAC of the message using K

$$\text{ActivePacket} = Ci\ S\ T\ P\ \text{Payload}\ \text{hmac}_K[Ci\ S\ T\ P\ \text{Payload}]$$

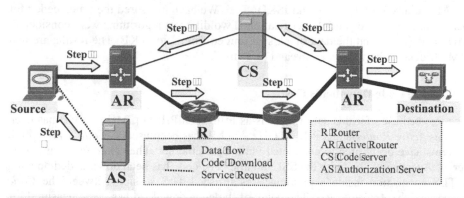

**Figure** The diferent steps

When an *Active Router* receives the message it first verifies that the message is not obsolete i e it is within the validity period and then it verifies the solicited active code availability In case the code and Kci is not locally available it downloads it using a secure TLS connection from the *Code Server* step in figure Then the *Active Router* generates K using Ci S T and P extracted from the active packet and Kci obtained from the *Code Server* when the code was downloaded If the HMAC is verified it means that *Source* has been authorized to execute the requested code so the *Active Router* processes the packet using the requested code and forwards it to the next hop The same procedure is repeated on every *Active Router* along the path until the packet reaches *Destination* The subsequent active packets of the flow will benefit from cached copies of the active code and Kci in every *Active Router* It must be noted that the presented solution is limited to one security domain i e one *Authorization Server* providing keys It is possible to extend the solution to multiple domains but this is more than a trivial task and it will be presented in future works

# Implementation and Tests

In order to evaluate the viability of ROSA, the security architecture has been implemented and its influence on the end to end delay has been measured. Two main processes have been evaluated: active packets protection and secure active code downloads. We have not considered the service request phase, because it has a similar cost to a code download and it is only executed once. In order to enable a simple integration of the developed prototype with available active network platforms, this implementation has been developed in Java, even though we are fully aware of the performance penalty of this choice.

## Active Packets Protection Cost

In ROSA, active packets protection is provided by HMAC, so performance of HMAC Java implementation has been measured. Tests have been done using a PIII Ghz, MB Linux Kernel and JSKD. We have measured the time needed for performing an HMAC and its verification. Two different algorithms were considered: MD and SHA, and the data block size ranged from to KB. The results are that HMAC delay is between ms and ms.

## Secure Code Downloads Cost

The test bed used for this set of trials is as follows: a PIII MHz MB has been used as *Active Router* and a PIII GHz MB has been used as *Code Server*, both systems have been directly connected via a Fast Ethernet. The *Code Server* is a web server Apache v with the SSL module. The code has been downloaded opening a TLS connection inside a previously established TLS session between the *Code Server* and the *Active Router*. The delay of the code download has been measured for non secure connections (http) and secure connections (https). Different code sizes from KB to KB have been used in the tests. The obtained delay is ms for http and ms for https. In our scenario, the *Code Server* and the *Active Router* will be in the same domain so we estimate an additional delay of ms, which would be the mean delay introduced by hops. Hence, for instance the estimated delay for KB code size would be ms for http and ms for https.

## Security Cost of an End to End Typical Communication

In this section, we will evaluate the cost of providing security to the active network inside a typical Internet scenario, composed of routers, four of which are *Active Routers*, with an average end to end delay of ms. We will state the following additional suppositions: average packet size B, average active code size KB, and all *Active Routers* modify active packets, so they must compute HMAC twice, one time to verify the received packet and another time to send the packet. Obviously, the active packet modification delay will be the same for the secure or non secure solution and therefore it will not be considered in the end to end delay.

Data obtained from http www caida org

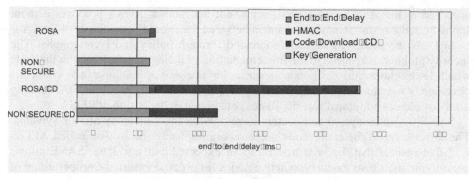

**Figure** Comparison of end to end delay

In a non secure scenario NON SECURE CD in figure the first active packet suffers a delay of ms ms end to end delay plus ms for code download Note that the final CD indicates that this packet has led each *Active Router* to do a code download In the ROSA scenario with code download ROSA CD in figure delay increases to ms because of https code download ms key generation in *Active Routers* ms HMAC in *Active Routers* *Source* and *Destination* ms

Next we will analyze the cost when the active code is already in the *Active Routers* which will be the situation in most cases The delay introduced by ROSA is of ms from ms in NON SECURE solution to ms because of the HMAC processed in *Active Routers* *Source* and *Destination*

The results show that ROSA introduces a small increase to the non secure end to end delay in most cases i e when code downloads are not needed Only the first active packet of the session experiences a higher delay We conclude that the delay introduced by ROSA is reasonable and the proposed architecture is feasible

## Related Work

In this section we will perform a comparative analysis of the proposed solution ROSA and other security architectures proposed for other platforms In particular we will study the security solutions presented for ANTS DAN and SANE

ANTS [ ] and ASP are the two main EEs running on Abone ASP specification does not define a security architecture but SANTS [ ] is an ANSA based proposal for ANTS ANSA is an Active Network Security Architecture ANSA [ ] proposed by the research community to be used in Abone [ ] ANSA uses symmetric key techniques over the variable part of the packet [ ] in order to provide inter node protection and digital signature over the fixed part of the packet to provide authentication and authorization of the principal In ROSA we avoid asymmetric key with the purpose of improving the performance That is possible because we do not need the non repudiation service over the active packets flow since we provide it during service request Furthermore since a topology independent solution is required the mechanism to share the keys based on the neighboring relationship proposed in the ANSA based EEs [ ] is not acceptable. ROSA proposes a more sophisticated key distribution mechanism between the trusted components of SARA.

In DAN the security issues are addressed through policy and cryptography. The security problem is reduced to the implementation of a simple policy rule on the node which lets it choose the right code server and a database of public keys to check the developer's signature of the plug in and the code server's authentication. DAN simply does not address additional security issues considered in the design of ROSA.

SANE [ ] is a layered architecture developed in the University of Pennsylvania. The lower layer of the architecture use a secure bootstrap mechanism called AEGIS [ ] that ensures that the system starts in an expected and safe state. SANE allows users to run their own modules on active nodes. In order to ensure the proper usage of network resources it authenticates and authorizes requesting users through the usage of a modified version of the Station To Station protocol (STS [ ]) between the user and each active node along the path that packets will follow. Once the STS protocol has concluded and a security association is established between the user and each active node the user's packets can be authenticated. This scheme imposes the usage of a different authenticator for each active node in the path which must be carried in each active packet. The proposed solution for this issue is that a common secret key is distributed among every active node using the established security association. The main drawback detected is the time needed for path establishment, since a STS exchange and a secret key exchange are needed. Moreover, when the path changes, these operations must be performed over the new path. ROSA has a reduced path establishment time since only one key exchange with the *Code Server* is needed.

## Conclusions

We have presented a security solution for active network platforms that follow the discrete approach. Key features of the solution include. The solution performance is guaranteed by the usage of symmetric key cryptography. The scalability of the solution is assured by the authorization model based on credentials, and the key distribution mechanism that minimizes key exchanges by allowing key generation at every *Active Router* in an autonomous fashion. The security level of the solution is determined by the re keying frequency i.e. how often Kci keys are changed. Essential features of the solution such as performance and scalability have been validated with measures obtained from a prototype implementation of the solution. Furthermore it must be stressed that the proposed architecture is open since it is valid for any active network platform as long it follows the discrete approach using a *Code Server*.

## Acknowledgments

The authors would like to thank Sandy Murphy for her review and comments. This work has been funded by CICYT under project AURAS.

# References

- Wetherall D J Legedza U Guttag J Introducing new Internet services Why and How IEEE Network Magazine
- Tennenhouse D L Wetherall D J Towards an Active Network Architecture Computer Communication Review Vol No April
- Decasper D Plattner B DAN Distributed Code Caching for Active Networks IEEE Infocom' San Francisco California March April
- Decasper D Parulkar G Choi S DeHart J Wolf T Plattner B A Scalable High Performance Active Network Node IEEE Network Jan Vol num pag
- Larrabeiti D Calderón M Azcorra A Urueña M A practical approach to network based processing $\text{th}$ International Workshop on Active Middleware Services July
- GCAP IST project home page http www laas fr GCAP
- SARA home site http enjambre it uc m es sara
- Berson S Braden B Ricciulli L Introduction to the Abone February
- Braden B Cerpa A Faber T Lindell B Pillips G Kann J Shenoy V Introduction to the ASP Execution Environment v November
- IST FAIN Project Deliverable "Initial Active Network and Active Node Architecture" Editor Spyros Denazis
- Krawczyk H Bellare M Canetti R HMAC Keyed Hashing for Message Authentication RFC April
- Dierks T Allen C The TLS protocol Version RFC January
- Wetherall D Guttag J Tennenhouse D L ANTS A Toolkit for Building and Dynamically Deploying Network Protocols Proceedings IEEE OPENARCH April
- AN Security Working Group Security Architecture for Active Nets November
- Murphy S Lewis E Puga R Watson R Yee R Strong Security for Active Networks Proceedings IEEE OPENARCH April
- Faber T Braden B Lindell B Berson S Bhaskar K Active Network Security for the ABone November
- Lindell B Protocol Specification for Hop By Hop Message Authentication and Integrity Dec
- Scott Alexander et al A Secure Active network architecture Realization in the SwitchWare IEEE Network May June vol no
- Arbaugh W et al Automated Recovery in a Secure Bootstrap Process Network and Distributed Systems Symposium Internet Society March
- Diffie W van Oorschot P Wiener M Authentication and Authenticated Key Exchanges Design Codes and Cryptography vol

# A Flexible Concast-Based Grouping Service

Amit Sehgal, Kenneth L. Calvert, and James Griffioen

Laboratory for Advanced Networking, University of Kentucky, Lexington, KY
{amit,calvert,griff}@netlab.uky.edu

**Abstract.** We present a scalable and flexible grouping service based on concast and best-effort single-source multicast. The service assigns participating end systems to specific groups based on application-supplied criteria. Example uses of such a service include peer-to-peer applications that want to group machines that are "near" each other, and reliable multicast services that need to assign receivers to repair groups. Our generic grouping framework relies on concast's many-to-one transport service to efficiently collect and apply the application-specific grouping criteria to the group members' information, and it relies on single-source multicast (i.e., one-to-many communication) to distribute the results to the nodes being grouped. The service can easily be customized to meet the grouping requirements of the application. We present simulation data showing the convergence properties of our grouping service and its effectiveness when applied to the problem of constructing overlay networks.

## 1 Introduction

Distributed applications often partition their member-set into groups to improve scalability and performance. In some cases this process occurs randomly, while in others it is controlled in order to ensure that nodes having similar characteristics end up in the same group. For example, the scalability of reliable multicast improves when receivers are grouped according to patterns of loss, and retransmissions are sent only to groups in which some member lost the packet [1, 2]. Peer-to-peer applications and overlays benefit from reduced latency when participants are grouped (connected to) others who are topologically "nearby".

Although examples of this kind of grouping abound, it is typically achieved through custom means; we know of no protocol designed to assist in group formation, say by assigning participants that satisfy the same application-supplied criterion to the same group. For small applications, a centralized approach to grouping is a viable solution: a single machine collects information from all participants, assigns them to their respective groups and returns the result to each participant. However, for large applications, the need to avoid implosion and reliably convey results to all participants limits the scalability of a centralized solution. Also, group formation decisions may be influenced by topology considerations. For example, the application may want to group nodes that are topologically "near" one another (for some definition of "near"). Thus, a grouping service must be able to take topology into account in some form.

In this paper we present a customizable *grouping service* that assigns network nodes to specific groups based on application-supplied criteria. The service uses single-source

J. Sterbenz et al. (Eds.): IWAN 2002, LNCS 2546, pp. 216–228, 2002.
© Springer-Verlag Berlin Heidelberg 2002

multicast to disseminate results, and concast to collect and process information from the participants. It is scalable because concast reduces the load on the central node and spreads the work of group formation across the network infrastructure. In addition, by processing information as it travels through the network, information about topology can be included in the process. The programmability of our service derives from that of concast, which allows an application to supply a *merge specification* describing how its packets should be combined as they travel through the network. Our grouping service is implemented as a generic merge specification plus an application-supplied data structure and three functions that operate on that data structure.

The remainder of this paper is organized as follows. Section 2 gives an overview of the concast service and Section 3 describes our grouping abstraction in detail. Section 4 demonstrates our service's applicability and applies the grouping abstraction to two problems: grouping receivers in reliable multicast and constructing topologically-efficient overlay networks. Section 5 discusses the simulation results that prove the convergence and scalability of our approach and Section 6 concludes the discussion.

## 2   Concast

This section provides a necessarily brief overview of the concast service and its programming interface. The interested reader is referred to [3] for a detailed description of concast.

A *concast flow* is uniquely identified by a pair $(G, R)$ where $R$ is the receiver's (unicast) IP address and $G$ is the *concast group identifier*, which represents the set of senders communicating with $R$. Concast packets are ordinary IP datagrams with $R$ in the destination field and $G$ in the IP options field (in a *Concast ID* option). The IP source address carries the unicast address of the last concast-capable router that processed the packet. Concast-capable routers intercept and process all packets that use the Concast ID option.

getTag($m$): A tag extraction function returning a hash or key identifying the message. Message $m$ and $m'$ are eligible for merging iff getTag($m$) = getTag($m'$)
merge ($s$, $m$, $f$): The function that combines messages together. The first parameter if the current merge state (i.e. information representing messages that have already been processed). The third parameter is a "flow state block" containing information about the concast flow to which $m$ belongs
done($s$): The forwarding predicate that checks s, the current merge state, and decides whether a message should be constructed (by calling buildMsg) and forwarded to the receiver.
buildMsg($s$): The message construction function, which takes the current message state $s$, and returns the payload to be forwarded toward the receiver.

**Fig. 1.** Merge Specification Methods

```
ProcessDatagram (Receiver R, Group G,
                 IPDatagram m) {
  FlowStateBlock fsb;
  DECTag t;
  MergeStateBlock s;

  fsb = LOOKUP_FLOW(R,G);
  if (fsb ≠ ⊥) {
    t = fsb.getTag(m);
    s = GET_MERGE_STATE(fsb, t);
    s = fsb.merge(s,m,fsb);
    if (fsb.done(s)) {
      (s,m) = fsb.buildMsg(s);
      FORWARD_DG(fsb, s, m);
    }
    PUT_MERGE_STATE (fsb, s, t);
  }
}
```

**Fig. 2.** The Concast Framework

The packets delivered to $R$ are a function of the packets sent by the members of $G$; the concast abstraction allows applications to customize the mapping from sent messages to delivered message(s), which is carried out hop-by-hop by the concast-capable routers along the path from senders to $R$. This mapping is called the *merge specification*; it controls (1) the relationship between the payloads of sent and received datagrams, (2) the conditions of message forwarding, and (3) packet identification (i.e. which packets are merged together). The merge specification is defined in terms of four methods (see Figure 1), which are invoked from a generic packet-processing loop as packets are processed hop-by-hop (see Figure 2).

The concast framework allows users to supply the definitions of these functions using a mobile-code-language. The merge specification functions are injected into the network by the receiver and pulled down into the network to the appropriate nodes by the *Concast Signalling Protocol* (CSP) when senders join the group.

Each concast-enabled router maintains the following information for each flow $(G, R)$ in a *flow state block*, or FSB.

**Merge Specification**:  the definitions of the **getTag**, **merge**, **done**, and **buildMsg** functions.

**Per-message State List**:  A list of in-progress "merge states" indexed by message tags.

**Upstream Neighbor List** (**UNL**):  Each item in the UNL represents a concast-capable node or a local sender (i.e. an application on the same node) for which the current node is the next concast-capable hop on the way to $R$. The node processes concast messages sent by the members of this list. Every member of this list is responsible for refreshing its state to avoid being purged from the list.

Incoming packets are classified into flows based on $(G, R)$. If no FSB is found for a packet, it is discarded. Otherwise the **getTag** function is obtained from the FSB and applied to the packet to obtain a tag that identifies the equivalence class of packets to which the packet belongs. The **merge** function is invoked on the current merge state for the tag (i.e., the merged state from all messages already received) to compute the new merge state. The **done** predicate then determines whether the merge operation is complete. If so, **buildMsg** is invoked to construct an outgoing message from the merged state that is forwarded toward the receiver.

CSP establishes concast-related state in network nodes. The protocol works by "pulling" the merge spec from the receiver towards senders as they join the group, installing concast state along the path from the sender to the receiver. All CSP message are sent as regular IP unicast messages with CSP identified in the protocol field, $(G, R)$ in the Concast ID option field, and the IP Router Alert Option [4] to stimulate hop-by-hop processing.

## 3    Grouping Abstraction

Our goal is a scalable service that can be used as a building block by applications that need to partition their members into groups. An easy solution is to support a predefined set of application-independent grouping policies. For example: network hop distance can be used as the group formation criteria where participants placed are placed in the same group if they are within a certain hop-distance of each other. However, predefined criteria will not suffice for many applications. Instead these applications want to form groups based on participant-supplied criteria. Our proposed application-specific grouping framework makes the following assumptions:

- All participants provide input (data) values (of a fixed, application-determined type) to the grouping service; this information is used to map participants to groups.
- There exists a mapping from the input values to a merged value of the same type that represents the group. (i.e. the values of the participants should be *merge-able* in some way other than simple concatenation.)
- All members of the application are multicast- and concast-capable.

Figure 3(a) illustrates the abstraction of the grouping service. One participant of the application initiates the service by providing the *grouping criteria*. The framework applies this grouping criteria to all participants' data in a distributed manner. Every participant $m$ of the application provides a value $v_m$ and an identifier $id_m$ as input to the grouping framework. The structure and content of $v$ and $id$ are defined by the application and interpreted *only* by the grouping criteria.

The framework, using the grouping criteria, compares and processes input values of all participants and assigns them to groups. The values of members that should be in the same group are mapped to a single group value $v_g$ that characterizes the members of the group. The framework nondeterministically chooses one of the group members' identifier as the group identifier $id_g$. Thus if $m'$, $m''$ and $m'''$ constitute the first group, then $\{v_{m'}, v_{m''}, v_{m'''}\} \rightarrow v_{g1}$ and $id_{g1} \in \{id_{m'}, id_{m''}, id_{m'''}\}$. Once completed, the grouping result is propagated back to all participants so that each can learn its own group value and the identifier associated with it. Thus from a participant's perspective, the grouping service takes an application-specific value and an identifier as input and in return provides the group value and the group identifier of the participant.

The grouping service provides one other configurable aspect: applications can choose different levels of reliability for the service. The strongest form of reliability requires the service to process the values of *every* participant before returning the grouping results. Another form collects input values of members over a specified interval of time and then propagates the results back, independent of the number of participants that have been grouped. In the following subsections we describe the implementation of our grouping service.

## 3.1 Grouping Algorithm

We use concast and single-source multicast to implement a scalable solution to the grouping problem. Figure 3(b) depicts the messages sent during the operation of the service. These messages are discussed in more detail in the following overview of the grouping algorithm:

1. One of the participants initiates the grouping service and provides the application-specific grouping criteria in the form of a merge specification to be used by concast. The initiator node functions as a concast receiver and a multicast sender in the grouping algorithm. As described in Section 2, the concast service deploys the mergespec (i.e. grouping criteria) in the network along the path from the members towards the initiator (concast receiver).

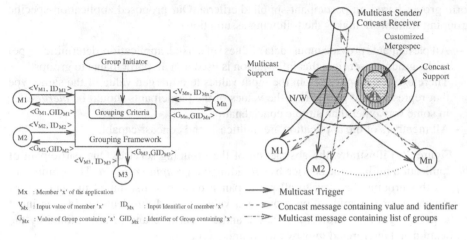

Mx : Member 'x' of the application

$V_{Mx}$ : Input value of member 'x'     $ID_{Mx}$ : Input Identifier of member 'x'

$G_{Mx}$ : Value of Group containing 'x'  $GID_{Mx}$ : Identifier of Group containing 'x'

→ Multicast Trigger

- - -▷ Concast message containing value and identifier

----▷ Multicast message containing list of groups

(a) Abstraction of the Grouping Service         (b) Messages used during Grouping

**Fig. 3.** The Grouping Service

2. The initiator multicasts a message that triggers all participants to respond.
3. Each member $m$ responds to the trigger with a concast message containing its application-specific value $v_m$ and identifier $id_m$.
4. Concast applies the grouping criteria to the concast messages. Each resulting group contains the merged value $v_g$ and a identifier $id_g$. A single message containing a (value, identifier) pair for each of the resulting groups is delivered to the initiator (concast receiver).
5. The initiator multicasts the list of groups back to all the participants.
6. Once a participant receives the entire list it identifies its group. We discuss the process of identification below.
7. The value and the identifier associated with the participant's group are ultimately returned to the application on each node.

Once a participant receives the list of $\langle v_g, id_g \rangle$ pairs, it needs to determine which group it belongs to. An easy solution is to have the grouping framework attach the unique node identifier[1] $nid$ of the member to the $\langle v, id \rangle$ pair input by it. When two values are classified under the same group, their corresponding $nid$'s can be concatenated to the $nidlist$ of the group. Thus the concast receiver would get a message containing a list of $\langle v_g, id_g \rangle$ pairs and a $nidlist_g$ associated with each pair. This message is multicast back to the participants who identify their $\langle v_g, id_g \rangle$ pair by searching for their $nid$ in the $nidlist_g$. However, this approach does not scale well, as the size of the $nidlist$ grows with the number of participants.

To achieve scalability, we propose a solution using Bloom filters. A Bloom filter [5] is a method for scalably representing a set of elements and then supporting membership queries for the elements. To summarize, a Bloom filter is a bitmap of $l$ bits and $k$ independent hash functions. For each output of the hash function (applied to some value $e$)

---

[1] This is not the same as the application-level identifier $id$.

the corresponding bit in the Bloom Filter bitmap is set. Thus bits $h_1(e), \ldots, h_k(e)$ are set in the bitmap. An element $x$ is assumed to be present if all the bits $h_1(x), \ldots, h_k(x)$ are set in the Bloom Filter. It is possible that a lookup finds an element, even though it was not inserted. Such a misidentification is termed a *false positive*.

At node $m$, the grouping framework hashes the unique node identifier $nid_m$ into an empty Bloom filter $bf_m$. The grouping framework couples $bf_m$ with the $\langle v_m, id_m \rangle$ pair and forwards them as part of the concast data. At the concast merging nodes, when two $v_m$'s belonging to the same group are merged, the corresponding $bf_m$ are combined by bitwise OR-ing them together. Thus, the amount of grouping information returned to participants depends on the number of groups and not on the number of participants.

The concast receiver finally receives a message containing a list of $\langle v_g, id_g \rangle$ pairs and their corresponding $bf_g$'s. This message is multicast back to the members, each of which performs a membership query for its own $nid$ on each group's Bloom filter. This query must succeed for at least one group. If it succeeds on exactly one Bloom filter the member has uniquely identified the group to which it belongs. However, membership queries may succeed on multiple Bloom filters due to false positives. To overcome this problem, additional iterations of the grouping algorithm are carried out, with one difference from the steps above: members that have already uniquely identified their groups do not hash their $nid$ in the Bloom filter in any of the subsequent iterations of the algorithm. This decreases the probability of false positives and helps the grouping algorithm converge faster. The grouping algorithm iterates until all members have uniquely identified their group, i.e. the concast receiver receives a message containing only empty Bloom filters.

## 3.2 Implementing the Grouping Algorithm

```
01 merge (MergeStateBlock msb, Packet p,
            FlowStateBlock fsb) {
02    for each record r in p
03        process_value (r.val);

04    if (msb is NULL)
05        msb ← create new merge state block
06        msb.seq ← p.seq
07        msb.neighbor_total ← fsb.unl_total
08        msb.neighbor_count ← 0
09        msb.timer ← NOW + δ

10    if (p.senderid ∈
            msb.neighbors_heard_from)
11        return (msb)

12    for each record r in p
13        for each record s in msb
14            if (can_combine (r.val, s.val))
15                s.val ← combine (r.val, s.val)
16                s.senderid ← choose (r.senderid,
                                        s.senderid)
17                s.bf ← r.bf || s.bf
18                delete r from p
19                break

20    for each record r in p
21        add r to msb

22    add (p.senderid to
            msb.neighbors_heard_from)
23    msb.neighbor_count ←
            msb.neighbor_count + 1
24    return (msb)
25 }

26 done_allnode (MergeStateBlock msb) {
27    return (msb.neighbor_count ==
            msb.neighbor_total)
28 }

29 done_timeout (MergeStateBlock msb) {
30    return ((msb.neighbor_count ==
            msb.neighbor_total) ||
31            (NOW ≥ msb.timer))
32 }
```

**Fig. 4.** Grouping Mergespec

Figure 4 describes the merge specification for implementing the grouping service. The initiator supplies three routines to the grouping service; *process_value* to pre-process the values of new incoming packets, *can_combine* to determine if two values are mergable and *combine* to merge two values. These three functions are used by the merge function (lines 1-25) at every concast node to achieve grouping. At the member node, the concast packet sent by the grouping framework carries the value $v$, the identifier $id$ and the Bloom filter $bf$ as data.

When a packet arrives at an intermediate concast node the merge function pre-processes it using *process_value*. If the merge state for the DEC corresponding to the incoming packet does not exist then it is created and all variables accordingly initialized. Next, if the incoming packet is a duplicate packet then the mergespec skips its processing. Otherwise *can_combine* determines if the records in the packet are mergable with the records in the merge state. If found to be mergable *combine* is used to merge the values in the records. **choose** non-deterministically picks either of the two $id$'s corresponding to the two merged values. Their corresponding Bloom filter's are merged using a simple *OR* operation. A merged record is deleted from the packet and in the end all unmerged records left in the packet are added to the merge state. The sender if of the processed packet is noted to avoid duplicate processing.

The mergespec also describes two kinds of **done** functions: **done_allnode** (lines 26-28) and **done_timeout** (lines 29-32). The two functions impose different conditions on packet forwarding at the concast node and as a result provide two different levels of grouping service. Either of the functions can be selected for use by the application at initiation time. Note that the application only needs to specify which of the two functions should be used and not provide the functions themselves.

One of the desired conditions of grouping is *complete reliability*, i.e. data from all participants must be processed to produce the final "merged" packet before delivering it to the concast receiver. This is accomplished by the **done_allnode** function which waits until it has received and processed a concast packet from all its upstream neighbors. But concast uses UDP for communication and packet losses during communication are possible. If the nodes wait indefinitely for the packets to arrive it may lead to deadlock. Thus if the concast receiver does not receive the merged concast packet, it times out and retransmits the trigger which causes the members to retransmit their $\langle v, id \rangle$ pair. Since the trigger is intended to complete the aborted process, it uses the same sequence number as the previous trigger. This redundancy ensures reliability.

The other desired functionality is that of *time bounded delivery*. This is accomplished by the **done_timeout** function which waits for a fixed duration of time after the arrival of the first packet and then forwards the merged packet towards the receiver. Thus members that fail to send their $\langle v, id \rangle$ pair within the fixed duration of time are excluded from the grouping process. This assures a time bounded response from the members once the trigger has been sent. In this case packet losses amount to excluding the member from the grouping algorithm.

The grouping process can be carried out periodically to incorporate new incoming members. However this can work well if the application has a low rate of participants leaving and joining and they are insensitive to the join delay. We are currently investigating techniques to make the grouping service adapt to more dynamic applications.

# 4  Applications

In this section we illustrate the use of our grouping service for two applications: reliable multicast and overlay networks. These applications use significantly different grouping criteria, demonstrating the flexibility of our grouping service.

## 4.1  Reliable Multicast

A common technique used in reliable multicast protocols—both router-based and end-system-based—is to assign receivers that need the same retransmission to the same channels (i.e. multicast groups) and thereby reduce redundant retransmissions.

A common technique is to use *lossprints* to record the retransmissions a machine requests [6, 7, 8]. Receivers with similar lossprints are likely to be "behind" the same set of lossy links with respect to the sender, and therefore can expect to request the same retransmissions in the future. Each multicast receiver keeps track of losses on the main multicast channel and generates a lossprint, which can be compared to the lossprints of other receivers to determine which receivers to combine into a retransmission group.

In end-system-based approaches, receiver lossprints must either be multicast to the entire group (thus creating additional load on the network in the form of bandwidth and routing state), or sent unicast to a central collection point (thus creating the risk of implosion). In our concast-based approach, receivers send their lossprints (in response to a trigger) toward the concast receiver. The lossprints are grouped, in the manner described below, hop by hop as they travel toward the concast receiver thereby reducing implosion and bandwidth usage. The concast receiver receives a single message containing a list of lossprints, each representing a group of receivers.

If losses occur *only* on bottleneck links lossprints accurately partition members into retransmission groups. However in practice, losses will occasionally occur on other links and the grouping algorithm must allow for a certain amount of "noise" in lossprints. Hence we combine two lossprints into the same group if they are within a threshold Hamming distance of their intersection (i.e. the bits that are set in both bitmaps). Each group is represented by the intersection of the lossprints of all of its members; this has the beneficial side effect of removing noise from the group lossprint as it moves up the tree.

Figure 5 illustrates the application supplied functions *process_value*, *can_combine*, and *combine* for this application. The application-specific data consists of a lossprint bitmap representing a fixed number of packets. The *process_value* function does nothing, as no pre-processing of the lossprints is required. The *can_combine* function compares two lossprints by first computing their intersection—a new lossprint that contains only the bits common to both lossprints. It then computes the Hamming distance (HD) between each lossprint and the intersection lossprint. The lossprints are considered combinable if both Hamming distances are less than $\varepsilon$. Finally, the *combine* function returns the intersection lossprint as the new group (lossprint).

## 4.2  Overlay Networks

Overlay networks offer advantages to applications that need special routing and addressing schemes. In particular, they allow unique routing and addressing schemes to

```
structure RM {
    bitmap lossprint
}

boolean can_combine (structure RM: s1,s2) {
    bitmap new_lossprint

    new_lossprint = s1.lossprint & s2.lossprint

    return ((no_of_bits (new_lossprint ⊕ s1.lossprint) < ε )
    && (no_of_bits (new_lossprint ⊕ s2.lossprint) < ε))
}

void process_value (structure RM: s1) {
    return
}

structure RM combine ( structure RM: s1, s2) {
    structure RM s3
    s3.lossprint ← s1.lossprint & s2.lossprint
    return (s3)
}
```

**Fig. 5.** Reliable Multicast Grouping Mergespec Functions

```
structure Ovly {
    integer hops
}

void process_value (structure Ovly s1) {
    s1.hops ← s1.hops + 1
}

boolean can_combine (structure Ovly: s1,s2) {
    return ((s1.hops < θ) && (s2.hops < θ))
}

structure Ovly combine (structure Ovly: s1,s2) {
    structure Ovly s3

    s3.hops ← max (s1.hops, s2.hops)
    return (s3)
}
```

**Fig. 6.** Overlay Mergespec functions

be implemented rapidly, without extra network support beyond standard services. Constructing an overlay requires some means of locating other participants and connecting with them to set up the overlay. In the recent past, peer-to-peer applications like Gnutella, CAN, and Chord [9, 10, 11] have proposed some interesting mechanisms for forming and using overlay networks. In this paper, we deal with the specific case of Gnutella.

Gnutella is a protocol for distributed search. Every participant is a client as well as a server. To join the Gnutella network, a new node connects to any existing member of the network and probes its neighbors within a specified overlay-hop radius to discover more members to connect to.

Because the overlay network is established at the application layer, the use of the underlying resources may not be optimum, i.e. multiple connections over the same network link. Since Gnutella forms connections by randomly connecting to a known node and then to other nodes that are within some distance of it, the resulting connections are independent of the underlying topology. To minimize link stress it is desirable to group members by topological vicinity, so that a node's neighbors in the overlay topology also tend to be close to it in the underlying topology. This is especially important in a Gnutella overlay, which relies on expanding ring searches to locate content. If nodes cache content and access patterns are consistent across a topological group, access latency should benefit from the increased efficiency of the overlay connectivity.

Our aim is to construct a topologically aware overlay network. The idea is to form groups by combining nodes within a certain hop-distance (radius) of each other. We also want to identify one node as the group representative. When all members of the group connect to their representative node, they will also discover the identity of the other nodes in the group.

Figure 6 describes the application-specific information used for grouping along with the *process_value*, *can_combine*, and *combine* functions. The information consists of a single integer: the hop count. The IP address of the member is used as the identifier in this application. The hop count in the value structure carried by a concast packet reflects the number of *concast hops* traversed so far.

Every member sets the hop count to zero when originating its value, and sets the identifier to its own IP address. At an intermediate concast node, the merge function combines values in the packet with the existing merge state as follows: ***process_value*** increments the hop count. The ***can_combine*** determines two values to be mergeable iff the hop count of both the values is less than the threshold $\theta$. If so, ***combine*** "merges" the two values by selecting the larger of the two hop counts. The grouping framework non-deterministically chooses either of the IP addresses as the identifier(representative node) for the new merged group. Consequently, all members that lie within a radius of $\theta$ concast hops of the intermediate node are merged together into the same group and the IP address of a representative node is associated with each group. This enforces the required distance relationship, in concast hops measured towards the concast receiver, among the members of the group. The resulting overlay network is closer to the underlying topology than a randomly connected network.

# 5  Simulations

## 5.1  Topology-Based Grouping

To demonstrate the topological benefits of our grouping algorithm we simulated the overlay grouping mechanism proposed in Section 4.2 to form an overlay and compare its characteristics with overlays formed using the Gnutella approach (described later). The overlay graphs are generated on a 2376 node GT-ITM transit-stub topology. The member nodes and the concast receiver node are randomly selected from the stub domains. The number of members in the overlay varies from 200 to 1400 in steps of 200, with each member having three or four neighbors. Our grouping algorithm places members within a radius of *three* concast hops in the same group.

Member identify their group and the corresponding representative node. They make connections to the representative nodes and also discover other nodes in the same group to connect to. A few connections are made to representative members of other groups to achieve global connectivity. In the Gnutella approach connections are made randomly to any member of the overlay.

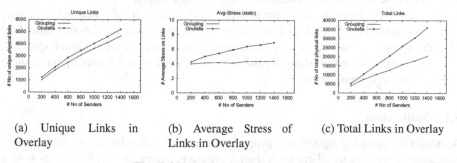

(a)  Unique  Links  in  Overlay

(b)  Average  Stress  of  Links in Overlay

(c) Total Links in Overlay

**Fig. 7.** Network Link Measurements

Connecitons formed using the grouping mechanism traverse fewer underlying network links as comapred to those using Gnutella, and this directly influences the number of links, the stress (defined as the number of connections that pass through the link) on those links, and the average hop distance to neighbors. These are important characteristics that affect latency and congestion issues. Figure 7 plots the network link characteristics of the two overlays and Figure 8 plots the connectivity characteristics of the two overlays.

Figure 7(a) illustrates that the number of unique links in the grouped overlay is marginally less than the unique links in Gnutella. However Figure 7(b) shows that the average stress on the links remains constant for grouped overlays across the number of receivers while the stress on the links in Gnutella increases with the increasing number of receivers. Figure 7(c) synthesizes the information in Figure 7(a) and Figure 7(b) by measuring the total number of links in the overlay and demonstrates the efficient use of links by the grouped overlays as compared to the Gnutella network.

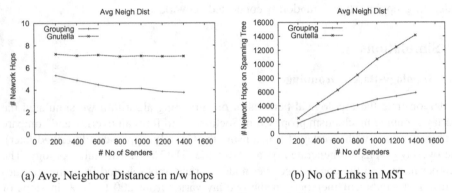

(a) Avg. Neighbor Distance in n/w hops      (b) No of Links in MST

**Fig. 8.** Overlay Characteristics

Figure 8(a) plots the average number of underlying network hops, taken over all pairs of neighbors, for both overlays. Since Gnutella connections are formed randomly, increase in membership does not affect the average network hop distance to the neighbors. In the grouped case, increase in membership casues the number of members within a group to increase. Thus more neighbors are now available to form connections and this reduces the average hop distance for the grouped overlay.

The average network hop distance to neighbors has a direct influence on the average delay experienced by the nodes when communicating with their neighbors. Figure 8(b) plots the average network hops in the connections that are part of the minimum spanning tree built using the two overlays. Again, we see that the grouped overlay has a more compact network connectivity.

## 5.2   Notification

Although the grouping algorithm is probabilistic in nature, it exhibits very nice convergence properties. As members learn their respective groups, they refrain from setting bits in the Bloom Filter which greatly reduces the false positive rate (i.e., the number of bits selected by more than one receiver). Even when the number of group members is significantly larger than the size of the bitmap in Bloom Filter (i.e., the false positive rate is extremely high – multiple members per bit), the algorithm converges in only a few iterations. The reason for this is that colliding members remain confused *only if they belong to different groups*. As long as the number of groups remains small compared to the number of members—which is arguably necessary for scalability of any grouping scheme—it is very likely that at least *some* colliding members will end up in the same group and become aware of their group despite false positives. This means that a fixed-size bitmap in the form of a Bloom Filter can be used with a very large member set.

To demonstrate the convergence properties of the grouping algorithm we simulated it on a 1000 node transit-stub network topology generated using the GT-ITM package [12] for the multicast grouping problem. The multicast members were randomly selected within the stub domains in the graph. We simulated our group formation policy described in Section 3 and Section 4.1 to group multicast receivers behind the same "congested" link into the same group. We randomly define $L$ links in the graph as "congested", resulting in at most $L + 1$ possible groups.

Figure 9a plots the number of confused receivers against the number of iterations needed to converge for 300 group members, 50 lossy links (i.e. 51 groups) and 3 different Bloom Filter sizes of 128, 256 and 512. The graph highlights the robustness of the grouping algorithm. Even with more than two receivers per bit, the algorithm has a steep convergence rate (i.e. converges quickly).

Figure 9b plots the average number of iterations needed to converge against the number of application members (receivers). In this case we varied the number of receivers from 100 to 800 in increments of 100 for Bloom Filter sizes ranging from 128 to 1024. The number of lossy links (groups) was always 10 percent of the number of receivers. This graph highlights the relative insensitivity of the algorithm to the number of receivers, for a constant receivers/groups ratio.

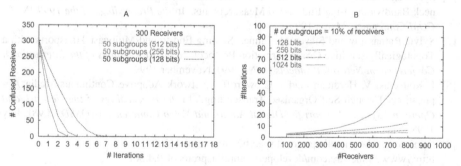

**Fig. 9.** Convergence rate for (a) different bitmask sizes, and (b) for differing numbers of receivers.

## 6   Conclusion

We have presented a scaleable and flexible grouping service based on concast and best-effort single-source multicast. The service does not require any router extension other than concast. Applications can make use of the service by supplying three simple functions to our generic group merge specification. These functions manipulate application-supplied data to implement application policy for partitioning members into groups. The functions are easy to specify, and can pose no security threat when used with the concast merging framework inside the network.

Concast makes it possible to collect the grouping information at a single point. Our announcement technique uses Bloom filters to enable each participant to learn its group assignment without explicitly identifying individuals in any message. Our simulation results show good convergence properties for our grouping service, even when the ratio of group size to Bloom filter size is large. Also, by customizing our grouping service to two diverse applications: reliable multicast and overlay networks, we show that the service is flexible and can be adapted to various applications. In the future, we plan to investigate how to apply the grouping framework to other classes of applications.

# References

[1] Sneha Kumar Kasera, Supratik Bhattacharyya, Mark Keaton, Diane Kiwior, Jim Kurose, Don Towsley, and Steve Zabele. Scalable Fair Reliable Multicast Using Active Services. *IEEE Network Magazine*, February 2000.

[2] R. Yavatkar, J. Griffioen, and M. Sudan. A Reliable Dissemination Protocol for Interactive Collaborative Applications. In *The Proceedings of the ACM Multimedia '95 Conference*, pages 333–344, November 1995.

[3] Billy C. Mullins Amit Sehgal Kenneth L. Calvert, James Griffioen and Su Wen. Concast : Design and implementaion of an active network service. *IEEE Journal on Selected Areas in Communications (2001)*, pages 19(3):426–437, March 2001.

[4] D. Katz. IP Router Alert Option, February 1997. RFC 2113.

[5] Burton Bloom. Space time tradeoffs in hash coding with allowable errors. *Communications of the ACM*, pages 13(7):422–426, July 1970.

[6] Sylvia Ratnasamy and Steven McCanne. Inference of Multicast Routing Trees and Bottleneck Bandwidths using End-to-end Measurements. In *the Proceedings of the 1999 INFOCOM Conference*, March 1999.

[7] Sylvia Ratnasamy and Steven McCanne. Scaling End-to-end Multicast Transports with a Topologically-sensitive Group Formation Protocol. In *the Proceedings of the International Conference on Network Protocols (ICNP '99)*, November 1999.

[8] I. Kouvelas, V. Hardman, and J. Crowcroft. Network Adaptive Continuous-Media Applications Through Self Organised Transcoding. In *the Proceedings of the Network and Operating Systems Support for Digital Audio and Video Conference (NOSSDAV 98)*, July 1998.

[9] Clip2.com. The gnutella protocol specification ver 0.4, 2000. http://www9.limewire.com/developer/ gnutella_protocol_0.4.pdf.

[10] M. Handley R. Karp S. Shenker S. Ratnasamy, P. Francis. A scalable content addressable network. In *in Proceedings of ACM Sigcomm '01 Conference*, August 2001.

[11] D. Karger F. Kasshoek I Stoica, R. Morris and H. Balakrishnan. Chord : A scalable peer-to-peer lookup sevice for internet applications. In *in Proceedings of ACM Sigcomm '01 Conference*, August 2001.

[12] Kenneth L. Calvert, Matthew B. Doar, and Ellen W. Zegura. Modeling Internet Topology. *IEEE Communications Magazine*, June 1997.

# Programmable Resource Discovery Using Peer-to-Peer Networks

Paul Smith, Steven Simpson, and David Hutchison

Computing Department
Lancaster University
Lancaster, LA1 4YR, UK
{p.smith, ss, dh}@comp.lancs.ac.uk

**Abstract.** Some forms of programmable networks such as funnelWeb
allow service components to be deployed discretely (i.e. out-of-band) on
a suitable configuration of elements, but do not define mechanisms to
determine such configurations.

We present a mechanism to resolve arbitrary service-specific deployment
constraints into a suitable node configuration. To focus constraint resolution, we arrange programmable elements into an overlay, and use this
to interpolate/extrapolate more favourable locations. Programmable service components are used to evaluate suitability of individual nodes.

## 1 Introduction

Critical to the successful operation of a network service is the deployment of
service components on appropriate network elements. The services supported by
a conventional network element are intrinsic to that element and static. However,
services on a programmable element are dynamic and transient. The challenge
in a programmable networking environment is to select from a set of *general
purpose* programmable elements a subset that is suitable for deploying a set of
service components.

When considering programmable service component deployment, two main
approaches have been adopted to date: the capsule (or in-band) approach, where
computational components are included with or referenced from within network
traffic that is to be augmented [WGT99, Wet99]; or the discrete (or out-of-band)
approach, where computational elements are loaded prior to the traffic's arrival
by a third-party process [FG98, SBSH01]. The discrete approach requires a process to select the appropriate set of programmable elements to deploy components on, unlike the capsule approach where component deployment is integral to
the manner in which the service is executed and so there is no such requirement.

In this paper, we are concerned with the discrete approach to service deployment. We present a critical part of a service deployment architecture that enables
the identification of programmable entities based upon a set of service-specific
constraints. Using our GROUPNET algorithm, programmable elements are organised into a peer-to-peer overlay network. Programmable service components are
used to resolve service-specific constraints into suitable node configurations.

J. Sterbenz et al. (Eds.): IWAN 2002, LNCS 2546, pp. 229–240, 2002.

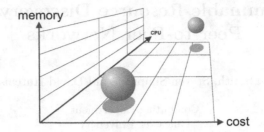

**Fig. 1.** An example set of dimensions and two different services with distinct desired values within those dimensions

We continue this section with a discussion of issues related to service deployment constraints. Following in Sect. 2 an overview of our GROUPNET overlay construction algorithm is given, which enables us to infer locations to search when resolving deployment constraints. Section 3 presents a search algorithm using GROUPNET and shows simulation results of a number of searches. In Sect. 4 we present an example service deployment scenario and show how our approach can be used. In Sect. 5 related work are discussed. Finally, in Sect. 6 we present some conclusions and outline further work.

### 1.1   Service Deployment Constraints

Peer-to-peer networking services such as Freenet [CSWH01] perform single-dimensional constraint resolution. For example, they transform the name of a shared file into a set of peers from which the file can be obtained. In programmable networks there are potentially multiple service deployment constraint dimensions that can be distinct on a per-service-instantiation basis, for example monetary cost and topological location. This is exemplified in Fig. 1, where one service has relatively low computational and cost constraints and another with relatively higher constraints.

In some cases, values within constraint dimensions can be traded against others in different dimensions – for example, a customer may be willing to pay a price in proportion to the service quality. Values within some dimensions can change in relation to network topology. For example, a telephone call may cost less when the end-points are topologically nearer. Programmable services that demonstrate these properties can be seen at [GF97, SMCH01].

A programmable element that is intended to support a number of potentially transient services cannot natively understand the nature of their operational constraints. It is for this reason we advocate the use of programmable service components to evaluate the suitability of individual elements. In this situation, elements expose interfaces that enable appropriately privileged programmable resource discovery components to interrogate local state and perform measurements with interesting nodes. Constraint resolution intelligence resides in programmable service components that interrogate *dumb* programmable elements.

## 2    GROUPNET: A GROUPing Meshed Overlay NETwork

Programmable elements are organised into an overlay network. This overlay network is constructed using the GROUPNET algorithm. A property of the overlay enables us to interpolate/extrapolate improved locations to search. A description of the GROUPNET overlay construction algorithm will now be given.

The procedure that a new peer wishing to join a GROUPNET mesh must follow can be described as follows. A peer $s$ locally decides its desired degree $d$, which will influence the number of immediate peers it has. It will maintain $N(s)$, its set of immediate neighbours. (Initially, using a bootstrap mechanism, $s$ is assigned a set of peers $N(s)$. This initial $N(s)$ can either be randomly selected, or to improve algorithm performance some heuristics can be used for selection.) $s$ now undergoes a sequence of optimisations.

In each optimisation, for each peer $p$ in $N(s)$, $s$ obtains measurements to $p$ and $N(p)$, and orders them to produce a list $L(p)$ (which includes $p$) of those peers in order. The new set of neighbours of $s$, $N'(s)$ will at least consist of the best of each $L(p)$ – if this falls short of $d$, the best of the union of the remaining members of all the $L(p)$'s can make up the shortfall.

The result is that for each $p \in N(s)$, $(N'(p) \cup p) \cap N'(s)$ is not empty, i.e. there is a path of no more than two hops between $s$ and each of its original neighbours, ensuring that the mesh does not become disconnected.

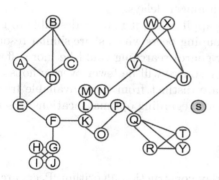

$d = 3$

1. $N(s) = \{K\}$

$N(K) = \{F, O, L\}$ , $L(K) = <O, L, K, F>$

2. $N'(s) = \{O, L, K\}$

$N(O) = \{K, P\}$ , $L(O) = <P, O, K>$

$N(K) = \{F, L, O\}$ , $L(K) = <O, L, K, F>$

$N(L) = \{K, P, M\}$ , $L(L) = <P, L, M, K>$

3. $N''(s) = \{P, O, L\}$

$N(P) = \{L, O, Q\}$ , $L(P) = <Q, P, O, L>$

$N(O) = \{K, P\}$ , $L(O) = <P, O, K>$

$N(L) = \{K, P, M\}$ , $L(L) = <P, L, M, K>$

4. $N'''(s) = \{Q, P, O\}$

**Fig. 2.** An example peer membership scenario with some of the stages of evaluation shown

## 2.1   Peer Membership Example

Figure 2 shows an example of a peer wishing to join a mesh and some of the steps toward achieving this. The node $s$ wishes to join the mesh and has selected a degree $d$ of 3. Using a bootstrap mechanism $s$ is assigned $\{K\}$ as its initial $N(s)$ as shown in step 1 of Fig. 2. The neighbours of $K$ are evaluated along with $K$ itself and the ordered list $L(K)$ is produced. In this example, peers closer to $s$ score higher than those more distant and therefore appear toward the start of the list. From $L(K)$, the best $d$ are taken to produce $N'(s)$ as shown in step 2. This completes a single iteration of the GROUPNET membership algorithm.

We anticipate a number of iterations would be performed until $N'(s)$ does not improve on $N(s)$. In step 2 of Fig. 2, $N'(s) = \{O, L, K\}$ which improves upon the immediate neighbours of $s$. Another iteration of the membership algorithm involves: determining $N(p)$ where $p$ is a peer in $N'(s)$ and $N(p)$ are all the neighbours of $p$ (for example $N(O) = \{K, P\}$), evaluating them to create $L(p)$ ($L(O) =< P, O, K >$) and then selecting the best from each $L(p)$ to generate $N''(s)$. If we find that we have already selected the best from a list, we simply take the next best as demonstrated in the example. It can be seen that the immediate peers of $s$ draw gradually closer to it. Ultimately, the algorithm will terminate with $N(s) = \{U, T, Y\}$.

## 2.2   Peer Evaluation Metrics

The metric used to evaluate a peer will depend on the application of the overlay mesh. For example, for programmable resource discovery we wish to construct a topologically aware overlay network. To this end, we could use round-trip delay between peers to evaluate suitability. Peers with smaller round trip delays score higher in this situation that those with longer delays.

If one considers a resource sharing application, it may be desirable to group peers sharing similar resources. By grouping peers which share similar resources (for example by music from a similar genre) searching could be more effective as peers probable to have a desirable resource will be fewer overlay hops away. If a user searches for resources that are distinct from those available from its immediate peers, inferences can be made regarding optimal locations to search.

## 2.3   Evaluation of GROUPNET

We have simulated the GROUPNET overlay construction algorithm. Peers are randomly distributed in a two-dimensional coordinate space and the group membership algorithm run at each peer. Distance between peers is used as the configuration metric. Peers with a smaller distance score better than those with larger distances. Figure 3 presents a graphical representation of the meshes produced. Node degrees of 3, 4 and 5 were assigned shown in Fig. 3a–c respectively. The top line of Fig. 3 (*random*) shows meshes produced after peers are assigned random neighbours. The bottom line of Fig. 3 (*groupnet*) show the meshes produced

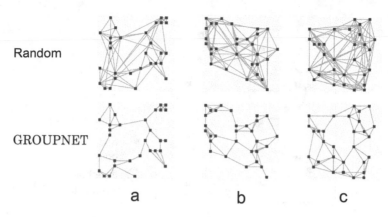

**Fig. 3.** Meshes created both randomly and using the GROUPNET algorithm with different node degrees

after running the GROUPNET overlay membership algorithm presented earlier. To optimise the mesh produced, we ran the algorithm 30 times at each node[1].

The aim of these simulations is to demonstrate that the membership algorithm produces overlays with desirable properties. The random meshes shown in Fig. 3 show undesirable properties as peers have links that are connected to relatively distant peers. The searching mechanism described in section 3 cannot be conducted on overlay configurations such as these, since there is no correlation between distances on the overlay and the space. The meshes constructed using the GROUPNET algorithm have the desired properties for enabling interpolation/extrapolation as there is a smooth transition over the space as one traverses the overlay.

## 3   Searching over GROUPNET

To evaluate the effectiveness and scalability of our approach, we have simulated searching over a GROUPNET mesh. To do this we construct meshes as in the simulations presented in Sect. 2. We define a target set of peers with three parameters: a point $p$ in the coordinate space, a threshold $T$ from $p$, that peers should fall within, and a timeout. Figure 4 shows the simulation scenario, with the target peers falling within the threshold shown.

A peer is capable of performing two operations on other peers. A node can either *evaluate* or *spawn* on another peer. The evaluate operation simulates remotely determining the suitability of another peer, for example by loading evaluation components. In our simulation, evaluation takes the form of a peer calculating its distance from $p$ and returning that value. The spawn operation

---

[1] Normally, we expect that optimisation would happen periodically, increasing in frequency the more excess links a peer has.

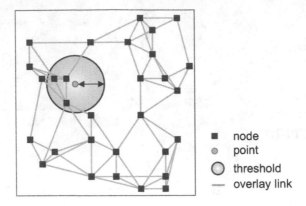

| | |
|---|---|
| ▪ | node |
| ⊙ | point |
| ◯ | threshold |
| — | overlay link |

**Fig. 4.** Scenario used for searching simulations

simulates instantiating programmable service components onto a peer. These are two operations we envisage as being necessary to conduct searches.

### 3.1  Search Algorithm

To conduct the search, a peer is randomly selected from the mesh. Initially, the timeout is checked and if it has expired the search algorithm terminates on that peer. The peer then randomly selects a number of other peers $n$ hops away and evaluates them. If the peers just evaluated have larger distances $s$ to $p$ than a peer's own $s$, a subsequent set of peers are selected $n/2$ hops away and evaluated. This process is iterated until $n = 0$, when the search is terminated at that peer. However, if a set of evaluated peers have smaller values of $s$ than a peer's own, the search algorithm is spawned on those peers. Peers with $s <= T$ are immediately reported after evaluation. The search should migrate or spawn toward peers with improved values of $s$ and terminate when better values of $s$ cannot be found.

### 3.2  Simulation Results

To assess the scalability of our search algorithm, we observed the average number of evaluations and spawns. We ran the search algorithm with the following parameters. We exponentially increased the number of peers in the mesh and adjusted the size of the coordinate space to maintain a constant density of peers. The value of $T$ was maintained to statistically yield a constant number of target peers. The initial hop count $n$ was kept at 20% of the mesh size. The algorithm was performed 50 times using each mesh size, with a new $p$ and initial peer to start the search on being selected at each iteration.

Spawns and evaluations are relatively expensive operations to perform. For example, programmable service component deployment using funnelWeb involves invoking a *load* command on a programmable element, fetching the component(s) over HTTP to the element and then instantiating the components. Figures 5 and 6 present the average number of spawns and the percentage of the total mesh spawned upon. The average number of spawns remains fairly constant in relation to increased topology size. The ability of the search algorithm to scale is best demonstrated in Fig. 6, where the total percent of the mesh spawned upon falls dramatically with the increase in size of the mesh.

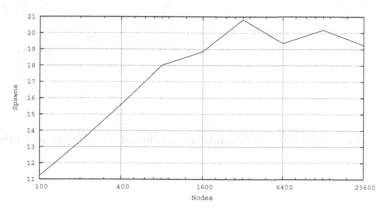

**Fig. 5.** Average number of spawns conducted on different size meshes

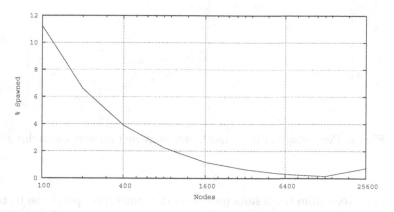

**Fig. 6.** Percentage of total mesh size spawned on when searching

Figure 7 presents the average number of evaluations conducted when executing the search algorithm. Again, as with the average number of spawns the number of evaluations remains relatively constant in relation to increased mesh size. The percentage of the mesh evaluated is presented in Fig. 8. The percentage of the total mesh evaluated falls significantly with increased mesh size.

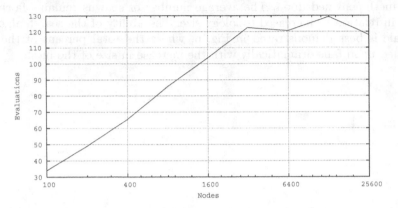

**Fig. 7.** Average number of evaluations conducted on different size meshes

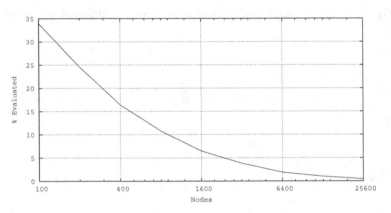

**Fig. 8.** Percentage of total mesh size spawned on when searching

It can be seen from the results presented that the target peers can be rapidly converged upon in a manner that scales with large mesh sizes. This is achieved by using a search algorithm that uses the properties of the GROUPNET overlay to infer improved locations to search. To demonstrate the effectiveness of the algorithm, we observed that out of the total number of peers to be found the algorithm located on average 89%.

# 4    Service Deployment Scenario

To demonstrate the applicability of our approach we will present an example service deployment scenario. The service we present in this example aims to reduce the cost of making long distance calls by using the Internet and Voice over IP (VoIP) to make the long distance connection [GFC00]. At the remote end of a connection, a network element with a modem attached to the PSTN can be used to complete a call to a local receiver, therefore making the total cost of a long distance call at most two local connections: one to connect to the Internet and the other to a local receiver. In our example scenario, the network element with modem functionality is programmable and the aim is to discover from the set of elements available a subset most suited to serve our application demands.

By using the Internet to form the long distance part of a connection, one exacerbates the problem of round-trip call delay. However the greater part the Internet constitutes the connection, the cheaper the cost of the call is likely to be. A balance must be struck between the total cost of the call and the delay experienced. Both parts of a connection contribute to a monetary cost and the total round-trip delay. We can use the function $c = c_n + c_p$ to calculate the total cost of making the call, where $c_n$ is the fixed cost of connecting to the Internet and $c_p$ is cost of the call on the PSTN. We use $t = t_n + t_p$ to determine the total round-trip delay, where $t_n$ is the delay across the Internet and $t_p$ is the delay across the PSTN. The values of $t$ and $c$ have acceptable upper bounds represented by $t_{max}$ and $c_{max}$ respectively. Elements with values greater than $t_{max}$ and $c_{max}$ can be immediately dismissed as unacceptable. The product of $t$ and $c$ can be used to select the most appropriate element from the available set. Table 1 shows example values of five candidate elements.

| Element | $t_n$ | $t_p$ | $c_p$ | $c_n$ | $t$ | $c$ | $tc$ |
|---------|-------|-------|-------|-------|-----|-----|------|
| 1 | 20 | 17 | 17 | 3 | 37 | 19 | 703 |
| 2 | 40 | 14 | 12 | 3 | 54 | 15 | 810 |
| 3 | 60 | 11 | 6 | 3 | 71 | 9 | 639 |
| 4 | 80 | 8 | 3 | 3 | 88 | 6 | 528 |
| 5 | 100 | 5 | 3 | 3 | 105 | 6 | 630 |

**Table 1.** Example costs for a set of candidate elements for deploying a Voice over IP service

If $t_{max} = 100\text{ms}$ and $c_{max} = 10\text{p/min}$ we can immediately consider elements 1, 2 and 5 from Table 1 as unsuitable. Elements 3 and 4 meet our constraints and we consider the value of $tc$ to select the most a appropriate element, in this case element 4 best meets the service constraints.

By constructing the overlay of programmable elements using GROUPNET, there is a correlation between relative location in the overlay and network topology.

One would anticipate the cost of the call $c_p$ to change in relation to network topology. Therefore traversing the overlay network should result in a consistent variation in the cost $c_p$. We can use this fact to interpolate/extrapolate better locations for searching. For example, if a direction along the overlay yields elements with a better $c_p$ than those already known, searching can continue in that direction in the hope that $c_p$ will continue to improve. Other inferences such as these can be made in order to find the set of programmable elements that have acceptable values of $c$. If such elements are found, $t_n$ can be determined and ultimately $tc$.

# 5   Related Work

There are peer-to-peer networking based solutions that are able to take multi-dimensional resource constraints and resolve/route them to the nearest set of peers that meet the constraints. Noteworthy approaches include CANs [RFH+01] and Pastry [RD01].

The design of Content-Addressable Networks (CANs) is focused on the use of a virtual $d$-dimensional Cartesian coordinate space, which is partitioned into zones that are controlled by peers in the CAN. A peer wishing to join the CAN selects a point in the coordinate space at which they wish to reside. They send a join message to an arbitrary point in the CAN with their desired coordinates. This message is routed over the coordinate space until it reaches a peer who controls the zone that occupies that point. The peer who owns the zone partitions it with the new peer and shares neighbour information. The new peer is now reachable across the CAN and is adjacent to peers with similar coordinates.

Peers in the Pastry network possess a unique node identifier that is obtained using a secure hash of its public key. As with CANs, a peer's identifier dictates its position within the Pastry network. The group membership algorithm used by Pastry enables peers to be co-located with others that have similar identifiers. Peers in Pastry maintain routing tables containing node identifiers of adjacent peers in the network.

Points in a CAN coordinate space and public keys in Pastry can be used to identify peers that satisfy multi-dimensional resource constraints. However, the values within a dimension are absolute – they do not vary in relation to different view points. It is also unclear how such approaches manage when there are highly dynamic values within dimensions as in a programmable networking environment.

For the purpose of programmable resource discovery, GROUPNET is used to construct a topologically-aware overlay network. Other approaches to constructing such overlays exist, for example Distributed Binning [RHKS02] and Hierarchical Clustering [MCSH02].

Distributed Binning involves associating nodes with logical *bins* based upon relative or absolute values that are derived from measuring a node's distance from a number of well-known landmarks. We believe that scalability issues could arise if a significant number of nodes measure themselves against a much smaller number of landmarks. More acutely, this problem could lead to a node being inaccurately associated with a bin due to significant link stress at a landmark.

Scalable Adaptive Hierarchical Clustering is a method of building a hierarchy of nodes, based on the notion of proximity, in a distributed and scalable way. The hierarchy is built through a series of "local" decisions involving only a small subset of the hierarchy's population for each decision. By using only a series local decisions and an innovative approach to adaptive cluster size distribution, the authors present a scalable means of constructing application-level overlay networks. Primarily this approach is intended for tree construction, however the algorithm can be trivially extended to generate overlay meshes.

## 6   Conclusions and Further Work

In this paper we have presented a novel approach to discovering a set appropriate programmable elements based upon service-specific deployment constraints.

We have suggested that deployment constraints differ in many dimensions on a per-service basis and programmable elements cannot natively understand the nature of these constraints. To solve this problem we propose that programmable service components are used to resolve deployment constraints and programmable elements make available to these components their local state for interrogation. We construct an overlay mesh consisting of programmable elements, which has the property of an element's immediate peers being topologically close. Using this overlay, we can perform searches on constraint dimensions whose ranges differ relative to network topology.

We have demonstrated some of the dimensions of scalability and effectiveness of searching over a GROUPNET mesh. Further work is required to evaluate the scalability and timeliness of the overlay construction algorithm. However, we feel this is secondary to searching performance, as joining the overlay would only occur at boot time. The use of programmable components for resolution may present unacceptable performance issues, we need to investigate these and suggest how performance could be improved. Additional consideration needs to be given to searching in multiple dimensions and how to converge on a configuration of resources with acceptable values within each. Further challenges lie in searching dimensions that have less correlation to the target and are not considered a part of the mesh dimensions.

## 7   Acknowledgements

This work was carried out under the Alpine project funded by BTexact Technologies. Paul Smith has a CASE studentship with the EPSRC and BTexact technologies.

## References

[CSWH01] I. Clarke, Oskar. S., B. Wiley, and T.W. Hong.  Freenet: A Distributed Anonymous Information Storage and Retrieval System. In H. Federrath, editor, *Designing Privacy Enhancing Technologies: International Workshop on Design Issues in Anonymity and Unobservability*, volume 2009 of *LNCS*, New York, 2001. Springer.

[FG98]    M. Fry and A. Ghosh. Application Level Active Networking. In *Computer Networks and ISDN Systems*, 1998.

[GF97]    A. Ghosh and M. Fry. Javaradio: an application level active network. In *Third International Workshop on High Performance Protocols (HIP-PARCH '97)*, London, June 1997.

[GFC00]   A. Ghosh, M. Fry, and J. Crowcroft. An Architecture for Application Layer Routing. In *IWAN 2000*, Tokyo, Japan, October 2000.

[MCSH02]  L. Mathy, R. Canonic, S. Simpson, and D. Hutchison. Scalable Adaptive Hierarchical Clustering. *IEEE Communications Letters*, 6(3):117–119, March 2002.

[RD01]    A. Rowstron and P. Drushcel. Pastry: Scalable, distributed object location and routing for large-scale peer-to-peer systems. In *IFIP/ACM Middleware 2001*, Heidelberg, Germany, November 2001.

[RFH⁺01]  S. Ratnasamy, P. Francis, M. Handley, R. Karp, and S Shenker. A Scalable Content-Addressable Network. In *ACM Sigcomm 2001*, August 2001.

[RHKS02]  S. Ratnasmay, M. Handley, R. Karp, and S. Shenker. Topologically-Aware Overlay Construction and Server Selection. In *IEEE INFOCOM 2002*, June 2002.

[SBSH01]  S. Simpson, M. Banfield, P. Smith, and D. Hutchison. Component Selection for Heterogeneous Active Networking. volume 2207 of *LNCS*, pages 84–100. Springer, September/October 2001.

[SMCH01]  P. Smith, L. Mathy, R. Canonico, and D. Hutchison. ALM and ProgNets for v4-to-v6 Multicast Transition. In *IEEE OpenArch 2001 - Short Paper Session "Ghosts of the Net!"*, Anchorage, Alaska, April 2001.

[Wet99]   D. Wetherall. Active network vision and reality: lessons from a capsule-based system. In *Operating Systems Review*, volume 34(5), pages 64–79, December 1999.

[WGT99]   D. Wetherall, J. Guttag, and D. Tennenhouse. ANTS: Network Services Without the Red Tape. *IEEE Computer*, 32(4):42–49, April 1999.

# Feature Interaction Detection in Active Networks

Go Ogose, Jyunya Yoshida, Tae Yoneda, and Tadashi Ohta

Soka University, Faculty of Engineering
1-236, Tangi-cho, Hachioji-shi, Tokyo 192-8577, Japan
{gogose, jyoshida}@edu.t.soka.ac.jp
{anne, ohta}@t.soka.ac.jp

**Abstract.** This paper presents the results of experiments with an active gatekeeper for VoIP. Problems arising with supplementary service programs uploaded by users: such as event conversion and detection of feature interactions between the programs, are presented and their solutions are proposed. The VoIP gatekeeper (GK) and the validation server, based on architecture, proposed by the authors at IWAN2000 and called STAR, were implemented and evaluated. With the proposed GK, the uploaded program is described using a declarative language ESTR instead of using a procedural language, such as Java. Unlike most conventional architectures, here, one common execution environment is used for execution with all up-loaded programs. 12 service programs were tested to evaluate the proposed systems. Results show that the proposed GK and validation server were reasonable. ...

## 1 Introduction

VoIP is becoming a standard technique to provide telecommunication services in the IP network. A VoIP gatekeeper has functions such as address translation from a telephone number to an IP address, network access control, and so on. Many architectures for Active Network have been proposed in order to provide new services instantly [1][2]. The authors are investigating using VoIP gatekeeper as an active node. The active VoIP gatekeeper (GK) and the validation server based on the architecture proposed by the authors at IWAN2000 [3] and called STAR (Software archiTecture for Active network using Rule based language), were implemented and evaluated. This paper presents results of experiments with STAR. Problems arising with supplementary service programs uploaded by unspecified users (third party service providers and end users): such as event conversion and detection of feature interactions between programs, are presented and their solutions are proposed. The problem of feature interaction, in particular, is a serious problem in developing an active network.

With the proposed GK, the up-loaded program, from now on referred to as the 'service program', is described using a declarative language ESTR (Enhanced State Transition Rule) instead of using a procedural language, such as Java. ESTR makes it easier to detect feature interactions. Unlike most conventional

J. Sterbenz et al. (Eds.): IWAN 2002, LNCS 2546, pp. 241–252, 2002.

architectures, here, one common execution environment is used for execution with all service programs.

12 service programs, including POTS (Plain Old Telephone Service), were described using ESTR and validated on the validation server. Because of the limitations of the system purchased, 11 service programs were tested on the proposed GK. TWC (Three Way Call service program) was tested using a simulator. It was confirmed that all service programs were executed correctly.

The results of experiments were evaluated by comparison with the international benchmark [4][5]. It was confirmed that all interactions described in the benchmark were detected. In fact, many more interactions, which were not described in the benchmark, were detected.

Results show that the proposed GK and validation server were reasonable.

Section 2 contains a brief description of the proposed architecture, STAR. In section 3, problems with implementation of the GK and the validation server are described. In section 4, solutions for the problems are proposed. Section 5 evaluates the proposed systems.

## 2   STAR

Much research into Active Networks has been done all over the world. In the IWAN2000, we proposed an architecture for the Active Network for VoIP Gate Way where the up-loaded program, ('service program'), is described using a declarative language. We have also implemented an experimental VoIP gate-keeper based on the proposed architecture (Figure 1). First, the proposed architecture STAR is explained.

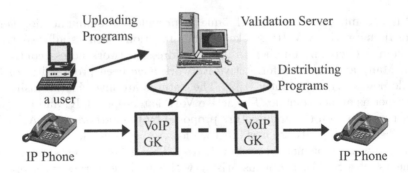

**Fig. 1.** Proposed Active Network

Characteristics of STAR are as follows:

1) A declarative language, ESTR, is adopted to describe service programs instead of a procedural language, such as Java.

**2)** One common Execution Environment (EE) is used for all service programs, instead of using an individual EE for each service program.

**3)** A Validation Server is used to detect feature interactions between service programs described by unspecified users, before the service programs are installed to network nodes.

These characteristics are explained briefly.

## 2.1  Related Work

Several special languages to describe service programs for the Active Network has been reported [6]. But, no declarative language other than ours has been reported as description language for the service program in the Active Network, especially for an active gatekeeper. Feature interaction detection is one of the most exciting research themes in the field of telecommunication systems research. Much research for definition, detection, and resolution of feature interactions has been discussed, mainly in the International workshop on feature interaction (FIW) [7-11]. But, as far as the authors know, there has been no proposal for feature interaction detection in the Active Network.

## 2.2  ESTR

The minimum explanation of ESTR, which is necessary for this paper, is given. See paper [2] for a detailed description.

ESTR has the form of Pre-condition, event, Post-condition and Action-description. It is a rule which defines the condition for state transition, state change while the rule is applied, and the system control required for state transition. Pre-condition consists of states description elements called primitives. Primitives are states of terminals or the relationship between terminals that are targets of the state transition. An event is a trigger that causes the state transition, e.g. a signal input to the node and some trigger occurs in the node. Post-condition is the state description part that also consists of primitives. Action-description is the system control description part that shows the system controls required for the state transition. Action-description is described in  which follows after Post-condition separated by ', '(see Figure 2). When no system controls are required, the content of  is empty. A description example of ESTR is shown in Figure 2.

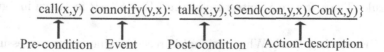

Pre-condition  Event    Post-condition   Action-description

**Fig. 2.** An Example of ESTR

The example in Figure 2 is explained. Terminal x and y are in calling state, denoted by call(x,y). If terminal y makes off hook, denoted by connotify(y,x), a signal Connect is sent to terminal x, denoted by Send(con,y,x), and terminal x and y transit to talk state, denoted by talk(x,y). call(x,y) and talk(x,y) are called status primitives. All arguments in status primitives are described as variables so that a rule can be applied to any terminals.

When an event occurs, a rule which has the same event, and whose Pre-condition is included in the system state, is applied. When the rule is applied, stored programs designated by Action-description are executed. When the programs end normally, the system state changes as follows. A state corresponding to the Pre-condition of the applied rule is deleted from the current system state and a state corresponding to Post-condition of the applied rule is added. Here, a state corresponding Pre/Post-condition is obtained by replacing arguments in Pre/Post-condition with actual terminals when the rule is applied.

## 2.3  Execution Environment (EE)

In most conventional architectures for the Active Network, each EE for the service program is uploaded to network nodes with the respective service programs [1]. Consequently, uploaded programs become very large. This adds significantly to the workload of users wanting to upload service programs. However, with STAR, one common EE is used for all service programs. Thus, users simply have to upload service programs only. Comparison of the software architectures of STAR and conventional architectures is shown in figure 3.

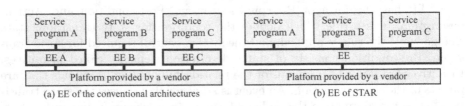

(a) EE of the conventional architectures        (b) EE of STAR

**Fig. 3.** EE Comparison

For STAR, the EE consists of: 1) an ESTR Interpreter, which selects a rule and executes the selected rule, 2) an Input processing part, which receives input signals from a platform provided by a vendor and converts them into events, and 3) an Action executing part, which analyzes and executes the Action-description of the rule. The software architecture of the EE of STAR is shown in figure 4.

**Input Processing Part** When a signal is received, the Input processing part converts the signal to an event corresponding to the signal, so that the ESTR Interpreter can handle the event. The event is sent to the interpreter in the EE program.

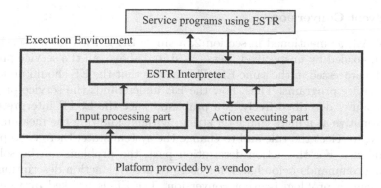

**Fig. 4.** Software Structure of EE

**ESTR Interpreter**  On receiving an event from the Input processing part, the interpreter selects a rule which has the same event as the one sent.

When the rule to be applied is selected, the Action-description of the rule is sent to the Action executing part in the EE program. If execution in the Action executing part ends normally, the system state is changed according to the Post-condition of the selected rule. If execution in the action executing part ends abnormally, the system state is not changed.

**Action Executing Part**  On receiving an Action-description from the interpreter, the Action executing part of the program analyzes it and decides which programs are to be executed and in what order. The programs will have been stored in the system beforehand by the service provider.

### 2.4  Validation Server

In an Active Network, unspecified numbers of users send their service programs to the nodes. This may cause feature interactions between the service programs. Feature interactions may cause serious problems to the network. Therefore, before being installed to the network nodes, the service programs are tested at the validation server. Multiple validation servers are set in the network and each validation server sends the service programs to nodes to which the user wants to upload the programs.

## 3  Problems with Implementing the System

The gatekeeper (GK) was implemented based on the proposed architecture [3]. Some problems with implementing the GK are described.

## 3.1   Event Conversion

With STAR, as mentioned in section 2, a single EE is used for all service programs up-loaded by unspecified users. As stated above, all the service programs should be processed in the same EE. This means that the EE should understand all the service programs. Thus, how the EE understands the service programs, which the user described freely, is a problem. Since the ESTR interpreter handles a primitive as a mere character string, without knowing the meaning of the primitive, to select the rule and to change the system state, there is no problem for primitives. For the action description part, the problem can be solved by preparing commands beforehand, for users to describe action description parts. The remaining problem is event conversion. The EE is invoked by receiving a signal. The ESTR interpreter is invoked by the event. Therefore, in order to invoke an interpreter, the signal has to be converted into an event.

On receiving a signal, the Input processing part converts the signal into an event described in the service programs. The signal event conversion is done as follows: the Input processing part searches an event conversion table, and converts the signal into the corresponding event.

The problem, therefore, is how to automatically revise the event conversion table when a new service program, which uses new events, is installed into the node.

## 3.2   Validation Server

Services that independently operate normally will behave undesirably when simultaneously initiated with another service. This behavior is called a feature interaction. As shown in AIN, JAIN, Parlay and Active Networks, a telecommunication network architecture changes to a new one where the third-party service providers can provide network services. This architecture enables multiple providers to provide services in the same network, simultaneously. As a result, feature interactions between different provider services inevitably occur. This causes serious problems to service deployment. A great deal of research on detecting feature interactions has been done all over the world [7-17] to solve the problems.

## 4   Solutions

### 4.1   Event Conversion

An event conversion table is used to convert a signal to an event. The event conversion table provides the relationship between an input signal and a corresponding event. To revise the event conversion table for newly used events, the following method is proposed:

When a user plans to use a new event in a new service program, he/she should send information on the correspondence relation between a new event and a signal with the new service to the validation server. After checking feature interactions, the validation server sends the information to the GK with the new service program. On receiving the information, the GK registers the information into the event conversion table, so that the Input processing part can identify the correspondence between a new event and a signal. On receiving a signal, the Input processing part converts the received signal into a corresponding event by referring to the event conversion table.

As an example of a registration process of a new event and an event conversion, a registration process and an event conversion of a specific number are explained. A specific number is a special telephone number for identifying a specific service.

Suppose, sp1setup(x) and 1901 are a newly defined event used in a service program and specific number input by an end user, respectively. First, the user, who uploads the service program which uses a specific number as an event, sends the information which defines the correspondence relation between the new event sp1setup(x), and the specific number, 1901, beforehand, to the validation server (VS). On receiving the information, the GK puts 1901 and sp1setup(x) into the event conversion table (Figure 5).

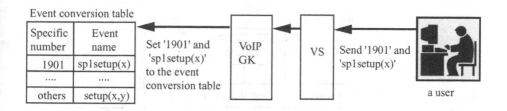

**Fig. 5.** Event Conversion Table Set

Next, an event conversion process for a specific number is explained (Figure 6). When the Input processing part receives a SETUP signal, the Input processing part calls a number analysis program. The number analysis program decides whether the terminating telephone number contained in a SETUP signal is a specific number or a usual telephone number, by referring to the event conversion table. If the terminating telephone number is a specific number, the number analysis program returns the event sp1setup(T1) to the Input processing part. If the terminating telephone number is a usual telephone number, the number analysis program returns the event setup(T1,T2) to the Input processing part. (Here, T1 and T2 are the originating terminal identifier and the terminating terminal identifier, respectively.)

## 4.2 Validation Server

**Outline** The Authors have proposed efficient methods for detecting feature interactions by analyzing Pre-conditions and Post-conditions of uploaded ESTR

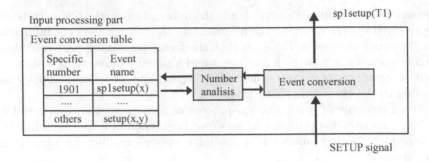

**Fig. 6.** Event Conversion Table Use

rules. To detect feature interactions between service programs uploaded by un-specified users, a validation server based on the proposed methods was implemented (Figure 7).

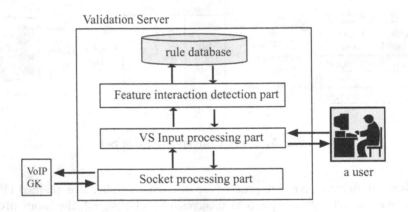

**Fig. 7.** Validation Server

The validation server checks for each uploaded ESTR rule whether the feature interactions are caused with existing rules. Feature interactions are classified into 7 categories [14]. The validation server implemented here detected the following 5 categories, which frequently occur in telecommunication services:

- Non-determinacy: there is more than one next state for one event.
- Appearance of abnormal states: Appearance of states which are not defined in the individual service.
- Disappearance of normal states: Disappearance of states which are defined in an individual service.

- Appearance of abnormal transitions: Appearance of transitions, which are not defined in an individual service.
- Disappearance of normal transitions: Disappearance of transitions, which are defined in an individual service.

First one causes EE service cancellation. Resting 4 interactions, called 'semantic interaction', cause various service malfunctions. When a feature interaction is detected between service A and B, the validation server regulates the service activation so that either of service A or B can be activated simultaneously.

**Detection Algorithm** A brief explanation of detection algorithm for semantic interactions will be described. For detailed descriptions, please refer to paper [13], [15], [16], and [17].

Semantic interactions can be considered as follows. Suppose two services are activated. When either specification of the services is applied, one state transition, according to the specification, clashes with the specification for the other service. Therefore, feature interactions are detected as follows: To make a rule pair, select a rule from each service, respectively, which is applicable to the same system state. Apply either rule to the system state. Check if the state transition by the rule causes an abnormal state transition from the viewpoint of the other service whose rule is not applied.

In some conventional methods, all possible states must be generated in one way or another and all state transitions checked to detect feature interactions. This causes an explosion of the number of states, resulting in a huge increase in computation time for detecting feature interactions.

In our method, on the other hand, interactions are detected solely by analyzing Pre-conditions, events, and Post-conditions of selected rules as follows:

**step 1)** If the rule pair can be applied to the same system state, go to step 2. Otherwise, a feature interaction is not detected.

**step 2)** If both rules have the same event, go to step 4, otherwise, go to step 3.

**step 3)** If the Pre-condition of the rule, which is not applied, is not preserved in the next system state, a feature interaction is detected. Otherwise, a feature interaction is not detected.

**step 4)** If the Post-condition of the rule, which is not applied, is not preserved in the next system state, a feature interaction is detected. Otherwise, a feature interaction is not detected.

Suppose, $r_a$ and $r_b$ denote selected rules from service a and service b, respectively. $r_{ac}$, $r_{an}$, $r_{bc}$, and $r_{bn}$ denote Pre-condition of $r_a$, Post-condition of $r_a$, Pre-condition of $r_b$, and Post condition of $r_b$, respectively. If redundancy can be neglected, formal descriptions of conditions used to detect interactions in step3 and step 4, respectively, are given as follows:

$$\text{For step3: } (r_{bc} - r_{ac}) \cup r_{an} \not\supseteq r_{bc}$$
$$\text{For step4: } \{(r_{bc} - r_{ac}) \cup r_{an} \not\supseteq r_{bn}\} \vee \{(r_{ac} - r_{an}) \not\supseteq (r_{bc} - r_{bn})\}$$

This method does not require any state creation and does not cause a huge increase in computation time for feature interaction detection. Other problems in implementing the feature interaction detection system are described in the next section. Redundancy will be evaluated in section 5.

**Detection Process** The detection process of the detecting system is shown in Figure 8.

First, select a pair of rules from the target service, one rule from each service, respectively [13]. After assigning real terminals to terminal variables in each rule [16], check if the pair of rules causes any non-determinacy interactions [17] and/or semantic interactions [13]. If the pair of rules causes any interactions, execute a reachability test. If the system state is reachable, the pair of rules actually causes interactions.

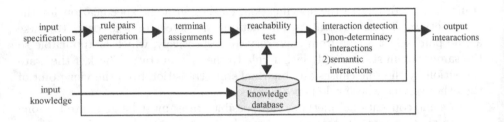

**Fig. 8.** Detection Process

## 5    Evaluations

A VoIP gatekeeper and a validation server were implemented based on the proposed methods mentioned in section 4. The gatekeeper was implemented by modifying an existing commercial gatekeeper.

### 5.1    Event Conversion

12 services programs: POTS, OCS(Originating Call Screening Service), TCS (Terminating Call Screening Service), CFV(Call Forwarding Service), FB(Free phone Billing), FR(Free phone Routing), CC(Charge Call service), CFBL(Call Forwarding Busy Line), TL(Teen Line service), TWC(Three Way Call Service), and two game service programs using specific numbers, a memory test game and a personality analysis game were described using ESTR. 8 services programs such as POTS, OCS, TCS, CFV, FB, FR, CC, and CFBL were uploaded to the GK. It was confirmed that all programs worked correctly. Although the two game service programs and TL were uploaded to the GK, it was only possible to

confirm event conversion functionality; this was because of the limitations of the commercially purchased gatekeeper and gateway. The same restrictions made it necessary to test the TWC using a simulator; results confirmed that it worked correctly.

Thus, it was confirmed that the proposed method of event conversion was reasonable.

## 5.2   Validation Server

Except for POTS, the 11 services mentioned in the previous section were validated on the implemented validation server. A benchmark for feature interaction detection was published in April 2000 [4][5]. To evaluate the proposed validation server, experimental results for nine service programs, which were shown in the benchmark, were compared with the benchmark. For example, the pairs of services of the feature interaction detected are TCS and CFBL, TCS and OCS, and CFV and FB. The feature interaction of TCS and CFBL is briefly explained. Terminal B activates CFBL, and registers terminal D as a forwarding terminal. Terminal D activates TCS, and registers terminal A as a screening terminal. Terminal A is off the hook, and terminal B is talking with terminal C. Then, if terminal A dials terminal B, and CFBL is applied, the state of terminal A will change the calling state with the terminal D. This is an illegal state transition because the system state doesn't change to a correctly state of TCS.

All the interactions described in the benchmark were detected, in fact, there are many more interactions detected, which were not described in the benchmark. It was confirmed manually that there was no redundancy and miss-detection.

Evaluation results for detecting interactions are given as follows:

1) A large number of interactions were detected, compared to what was shown in the benchmark. There was no redundancy or miss-detection. Thus it was confirmed that the proposed algorithm, including the detection system, is reasonable.
2) The reason so many interactions were detected is that, first of all, since ESTR allows a multiplicity of services to share states it has the mechanism to create feature interactions. With the conventional detection methods, limited terminal assignments were tested to suppress detection time because the correct way of making terminal assignments had not been made clear, but with the proposed methods, the way of making terminal assignments was clarified [16] and other filtering methods were developed. As the filtering methods reduced detection time considerably, all terminal assignments were considered and far more interactions were detected.

## 6   Conclusion

An active gatekeeper for VoIP and a validation server were proposed. Problems were identified and their solutions were proposed. The proposed systems were evaluated and it was confirmed that they were reasonable.

In future work, many more supplementary service programs will be applied in order to evaluate the proposed architecture and systems. Further work will be on how to guarantee that one user's action doesn't consume too many resources or disrupt other user's services. We are planning to make separations between users so that the problem does not occur.

# References

1. H. Yasuda Ed., "Active Networks," Lecture Notes in Computer Science 1942, Oct. 2000.
2. T. Morinaga, G. Ogose, and T. Ohta, "Active Networks for VoIP GW using Declarative Language," Proc. of APCC2001, pp.89-92, Sep. 2001.
3. S. Komatsu and T. Ohta, "Active Networks using Declarative Language," Proc. of IWAN2000, pp.33-44, Oct. 2000.
4. N. Griffeth et al., "Feature Interaction Detection Contest of the Fifth International Workshop on Feature Interactions," The Interanational Journal of Computer and Telecommunications Networking, Computer Networks 32 (2000) pp.487-510, April 2000
5. N. Griffeth et al., "A Feature Interaction Benchmark for the First Feature Interaction Detection Contest," The Interanational Journal of Computer and Telecommunications Networking, Computer Networks 32 (2000) pp.389-418, April 2000
6. K.L. Calvert et al., "Directions in Active Network," IEEE Com. Magazine, Vol.36, No.10, pp.72-78, Oct. 1998.
7. L.G. Bouma et al., Feature Interactions in Telecommunications Systems, IOS Press, 1994
8. K.E. Cheng and T. Ohta, Feature Interactions in Telecommunications III, IOS Press, 1995
9. P. Dini et al., Feature Interactions in Telecommunication Networks IV, IOS Press, 1997
10. K. Kimbler and L.G. Bouma, Feature Interactions in Telecommunications and Software Systems V, IOS Press, 1998
11. M. Calder and E.Magill, Feature Interactions in Telecommunications and Software Systems VI, IOS Press, 2000
12. T. Ohta et al., "Classification, Detection and Resolution of Service Interactions in Telecommunication Services," Proc. of FIW'94, pp60-72, May 1994
13. T. Yoneda and T. Ohta, "A Formal Approach for Definition and Detection of Feature Interactions," Proc. of FIW'98, pp.202-216, Sep. 1998.
14. T. Ohta and F. Cristian, "Formal Definitions of Feature Interactions in Telecommunications Software," IEICE transactions on Fundamentals of Electronics, Communications and Computer Science, vol.E81-A No.4, pp.635-638, April 1998
15. T. Yoneda and T. Ohta, "Automatic Elicitation of Knowledge for Detecting Feature Interactions in Telecommunication Services," IEICE transactions on information and systems, vol.E83-D No.4, pp.640-647, April 2000
16. T. Yoneda and T. Ohta, " Reduction of the Number of Terminal Assignments for Detecting Feature Interactions in Telecommunication Services," Proc. of ICECCS, pp.202-209, Sep. 2001.
17. J. Kobayashi, T. Yoneda and T. Ohta, "An Effective Method for Testing Reachability Using Knowledge in Detecting Non-Determinacy Feature Interactions," IEICE Trans. on Information and Systems, vol.E85-D No.4, pp.607-614, April 2002

# Flexible, Dynamic, and Scalable Service Composition for Active Routers

Stefan Schmid, Tim Chart, Manolis Sifalakis, Andrew C. Scott

Distributed Multimedia Research Group
Computing Department
Lancaster University, UK
{sschmid, chart, mjs, acs}@comp.lancs.ac.uk

**Abstract.** This paper describes a novel model for the provision of service composites for active routers. The service composition framework enables flexible
programmability of the router's data path through dynamically loadable software components called active components. The composition model promotes
transparent and dynamic creation of network side services and allows independent users to partake in this process. A prototype implementation has revealed that the composition model using packet filters and a classification graph
structure as a means to integrate active components into the forwarding path
enables the dynamic alteration of the elements of a composite at run time and
permits scalability in the generation of such composites. Furthermore, it allows
the flexible provision of a unique service profile for each packet passing
through an active router. We show that the overhead of this composition model
does not significantly affect the performance of the router.

## 1 Introduction

In recent years, many diverse and variously focused frameworks sporting elements of
active network solutions have emerged and established themselves. The majority of
those platforms address only a subset of the issues that determine their utility outside
of a laboratory environment, resulting in many being inflexible, poorly performing,
unscalable or insecure. Few active network solutions have considered the magnitude
of the service composition model for real life network environments and hence provide only a limited flexibility for the composition of network services. Yet, in order
for active networking to be considered a suitable technology for wide deployment
over inter networks, these issues must be addressed.

A recent study by Hicks and Nettles [1] has revealed that even most extensible
active router platforms lack sufficient flexibility in order to allow for true evolution.
Such modular or plug in based architectures typically limit the scope of future changes
through pre defined interfaces. Instead, true extensibility should not be limited to a
fixed set of modules or plug ins but should rather allow modification and replacement
of all components contributing to a service composite.

This paper presents the service composition framework – a key component of the
LARA++ [2] active router architecture – which attempts to provide a secure, safe,

J. Sterbenz et al. (Eds.): IWAN 2002, LNCS 2546, pp. 253–266, 2002.

flexibly extensible platform for the deployment of powerful active components. The main objective of the LARA architecture is flexible extensibility in order to facilitate network functionalities and services whose need may evolve in the future by making the entire forwarding path of the router programmable. The *packet classifier* is the LARA component responsible for the effective and flexible deployment of active components and the decentralised provision of an acceptable service composite.

The remainder of this paper is organised as follows. The second chapter gives a brief overview of the LARA architecture. Chapter continues with a detailed description of the service composition model. Chapters and describe the prototype implementation of the LARA classifier and the measurements made from that prototype. Chapter compares the composition model with other architectures. Finally, chapter concludes on our findings.

## Motivation

Before we describe the LARA architecture we outline some scenarios that adequately encompass many of the problems of service composition. Scenarios such as those described in this section precipitated the desire to provide a flexible and scalable composition model for active routers.

Suppose an active node has two components installed to process packets arriving on a given interface. One of the components might be a lightweight firewall component installed by the router administrator which is able to efficiently filter incoming packets to prevent processing of certain types of traffic. A second component installed by a network user might for example offer intelligent congestion control services for a custom protocol carried over UDP. The intelligent congestion control is likely to require much more processor time than the firewall so it would be advantageous for the firewall component to be processed first thus reducing the amount of traffic sent to the congestion control component. In order to ensure this optimisation, the active router must necessarily have a framework that allows an ordering of components to be maintained. Moreover, where two users dispute the ordering of components on the same stream user privileges must be taken into account.

Another example might involve a network user who would like to co-operate with active components already deployed on the active router. For example an existing active program on the active router might provide IPv transitioning support in the form of network address translation [ ]. IPv packets entering the router from one interface might be converted to IPv packets before being forwarded out of another and vice versa for the reverse path. This would enable IPv only hosts and IPv only hosts to converse. However this would cause any applications that embed IP addresses in the payload of the packet to 'break'. A network user wishing to provide support for such an application over a translating router would need to provide a component that co-operates with the network layer translation active component. The co-operation would require both components to be processed, but in a strict order. The challenge would be to find a way of asserting this co-operation over the same stream of packets without causing undesired interference between the components.

The problem of service composition is principally one of managing competition and co-operation between components in the processing of packets. The LARA service composition model described in this paper provides the structure required for a distributed (i.e. involving more than one entity) and dynamic (i.e. allowing service composition at runtime) composition model without restricting the flexibility and programmability of the active node.

## Background

LARA++ is a software architecture that evolved from the predominantly hardware oriented Lancaster Active Router Architecture (LARA) [ ]. The architecture is ge-neric in the sense that it can be implemented on top of any router platform with a software forwarding engine. Platform independence is assured by virtue of the fact that prototype implementations have been developed for Windows XP and Linux.

The LARA++ architecture is designed to extend existing routers with active net-work functionality. Low level functionality of the active router architecture is directly integrated with the router OS in order to maintain good performance for active proc-essing. A high level layer accommodates processing environments that enable safe processing of dynamically loadable software components. Well known interfaces are exposed at this layer in order to unify programmability across different platforms.

The processing environments provide the policing and code isolation necessary for a fair and safe platform. However since this paper focuses on the service composition model of the LARA++ architecture we refer the reader to previous publications [ ].

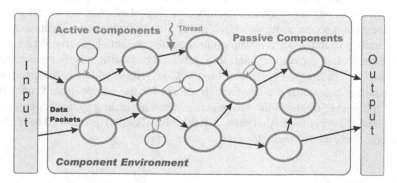

**Fig. . A Conceptual View of the Active Component Space.**

Figure provides a conceptual overview of the approach taken. The vision of the LARA++ platform is to provide a framework upon which complete router functional-ity can be provided in component form, for example a routing component, a filtering component, etc. which are then composed into actual network services at runtime. Componentisation has the advantages that it allows a "divide and conquer" approach to be employed for complex functionality, and that software components can be dy-namically extended and replaced due to their well defined interfaces.

---

Based on in band or out of band code loading techniques.

Service composition is achieved through packet classification. Packet filters define the processing "route" through the component environment. This allows such "routes" to be appropriately tailored to the packet type or content.

## LARA Composition Model

The LARA composition model plays a central role in the overall architecture, as it provides the foundation for the flexibly extensible and dynamic component based programming model. Service composition is carried out in two manners:

- *Macro* composition is achieved at the service level via the filter based composition model. Active components dynamically integrate themselves into node local service composites by inserting packet filters into the classification graph using a system call to the LARA NodeOS.

- *Micro* composition is achieved at the component level using an explicit lightweight composition model, which enables the construction of active components from passive components. The fact that active components can be largely composed from functionality provided by passive components facilitates active component design and co operation between active components (section).

Macro composition within LARA is largely packet driven. Dependent on the packet content a different overall service may be composed for the processing of that packet. The *packet classifier* plays a key role in the service composition process. It determines based upon the set of *packet filters* currently installed whether or not a packet passing through the active router requires active processing, which active component(s) are involved and in which order they should process the packet. Thus the active components implicitly and collectively define a service composite (resolving competition see section) in the process. The *classification graph*, which is managed by the packet classifier, maintains the key data structures for the composition framework. It organises the packet filters of the active components according to their computational function and thus provides the basis for the classification process. The following sections describe each of these elements in more detail.

### Packet Classifier

The packet classifier defines the "route" through the active component space for packets passing through a node. The classifier filters incoming and outgoing network packets based on the component filters installed in the classification graph. Figure presents an example classification graph.

The classification mechanism traverses the classification graph starting at a root node (ie `/netin` in figure). At each node in the graph the classifier tries to match the packet filters installed there. Packets matching a filter are passed to the corresponding active component. After completion of the active processing the classifier continues classification at the same point or an optionally specified point defined by the packet filter in the classification graph. When the classifier has applied all packet filters that have been installed by the active components at a node, it follows the classification graph based on the "default" or graph filters (for example, `/netin/ipv4` or `/netin/ipv6`) and continues the multi stage classification process there. Finally, the packet is forwarded to the next hop router when the classifier runs out of graph filters.

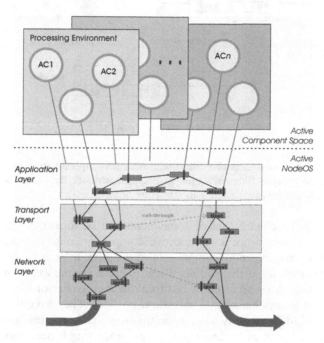

**Fig. The classification graph**

## Packet Filters

The packet classifier distinguishes two main types of packet filter, *active component filters* and *graph filters*. Figure illustrates how these filters are used within the classification graph.

Active component filters are used by active components to define the network packets of their interest. These filters are typically registered with the packet classifier at component instantiation or at run time if necessary, through the LARA system API. The classifier uses these filters to determine the active components to which network traffic is sent for active processing. For example, a customisable firewall component might register active component filters for each protocol the user has asked it to filter. Graph filters, in contrast, are used by the classifier itself in order to define the structure of the classification graph.

Availability of this option depends on the user privileges and filter type.

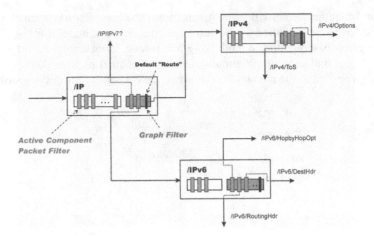

**Fig** The classifier manages packet filters within a filter graph structure called classification graph Active component filters are used to dispatch network data to ACs for processing whereas graph filters are used to define the structure of the graph

Packet filters installed by active components can be further divided into *general filters* and *flow filters* Since a single active component could install multiple filters and given there may be many components running on an active router it would be very costly to check all of those filters Flow filters a specialization of general active component filters have been introduced to allow the number of filters to scale well Flow filters are always bound to a specific user flow They have the advantage that they can be looked up instantly based upon the flow characteristics of the packet being processed using a hash table Consequently no processing is necessary to reject most unmatched flow filters Due to the hashing technique used to lookup flow filters a LARA router can handle a potentially large number of these filters which is fully sufficient for typical edge routers – the target platform for LARA

All types of filter define patterns against which the active node attempts to match characteristics of packets Such characteristics are commonly fields in packet headers and are described by a four tuple of {packet offset, bit pattern, bit mask, pattern length} The specification of packet filters is facilitated through well known reference points and pre defined tags For example a filter for HTTP traffic might make use of the TCP_HEADER reference point like so {TCP_HEADER + TCP_PORT, 0x0050[4], 0xffff, 2}. Packet filters are fixed in a single classification node known as the *filter input node* which must be appropriate to the filter For example the HTTP filter defined above could be placed in the classification node dedicated to TCP traffic because the filter requires TCP as a prerequisite Graph filters and some specially privileged active component filters also define a *filter output node* which provides an alternative route *cut through path*

---

A user or end to end flow is defined by the source destination or both end points An end point may be identified by packet fields such as the network layer addresses and transport layer ports or any other flow labeling techniques

x decimal is the TCP port to which HTTP traffic is directed

through the classification graph. For any packets matching those filters classification commences at the filter output node.

Active component filters require additional properties: (i) an *operation property* to express the packet access permissions required by the component (i.e. read only, read write or write only) and (ii) a *principal* security credential to indicate the network user on whose behalf the filter is installed and/or the code producer. The operation and principal properties permit the classifier to authorise the insertion of a filter based upon the node local security policy and the privileges associated with the principal.

## Classification Graph Table

The classification graph table (CGT) provides the means to describe the structure of the classification graph. Its main purpose is to make the graph structure globally available across a LARA++ active network. In order to support flexible extension of the classification graph (e.g. to incorporate new protocols or extend current protocols) an "elastic" means to describe the graph structure is required. For this purpose a simple notion for defining the nodes (for example `ipv4`, `tcp` and `udp`) and the branches of the graph (for example `ipv4→tcp` or `ipv4→udp`) has been introduced.

The basic structure of the classification graph described in the CGT conforms to the TCP/IP layer model, which ensures that active components providing low level services are processed before components dealing with higher level computations. For example, network protocol options must be processed prior to transport protocol headers. The fine grained structure accounts for the layer specific protocols. For example, extension headers in IPv6 must be processed in a pre-defined order.

The active node security policy specifies the types of filters that may be installed inside the classification nodes. For example, it would make no sense to allow general purpose filters to be processed before filters pertaining to a firewall component (or otherwise the security measure could possibly be circumvented). Classification nodes also define access permissions for fine grained control of packet access.

Since the CGT is expected to change occasionally (for example a new node in the global classification graph might be introduced when a new protocol becomes established), an automated mechanism to update the CGT across the active network is employed. At a first glance it may seem that the overhead of updating the CGT every time a new protocol is introduced is heavyweight and makes the system inflexible. However, it should be noted that a CGT update is only required if a new protocol or protocol extension is "standardised" (i.e. globally announced such that it can be extended by others). The CGT does not require a global update in order to deploy and test the new protocol or extension locally.

## Composite Characteristics

Service composition within LARA++ is a *co-operative* process. It allows independent network users to install active components that match the same data streams or subsets of streams (controlled by the local security policies). The classification graph provides the means for independent users to integrate new active functionality or services in a "meaningful" way without having to know about other users' active components. The decoupling of component bindings among active components through the classification graph hides component changes from other components.

The fact that active services are composed through insertion or removal of packet filters at run time, when components are instantiated or removed, makes the composition process highly *dynamic*. Since the service composite depends on the actual data in the packets, the component bindings are *conditional*. Service composition within LARA++ is therefore a process that takes place on a per packet basis.

## Implementation

This chapter outlines the implementation of the LARA++ composition model, and some of the features of the implementation that affect overall performance.

### Design Overview

The classification component of the LARA++ architecture is a key subsystem. Interactions between the classifier and the other components affect the overall performance of the implementation.

The classification component is responsible for the dispatch of classified packets to active components. When the classifier matches the filter of an active component, it inserts the packet into the packet channel (i.e. input queue) of the corresponding component and continues the classification process. This allows the classifier to efficiently classify packets without waiting for active components to be ready to receive them.

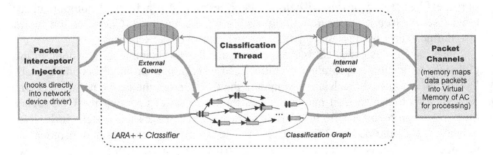

**Fig.** The LARA++ Classifier Architecture

Figure illustrates the architecture of the LARA++ classifier. Incoming packets, intercepted by the packet interceptor component, are asynchronously queued on a circular buffer known as the *external queue*. The classification thread sequentially takes packets from the external queue and performs an initial classification on them.

After a packet has been classified, i.e. a component has been selected to perform active processing on the packet, the packet is immediately queued in the packet channel of the active component, so that the classification engine can continue to classify more packets until all packets are classified or until its scheduling quantum is over.

Once classified packets have been processed by the corresponding active components, they are returned to a second queue known as the *internal queue*. The purpose of the internal queue is to hold packets prior to the continuation of their classification.

Many components will often be identified to process a packet over the course of its passage through the active router. However, it is not possible to identify all components to which a packet will be sent in advance as the components could change the content of the packet. Therefore re-classification of packets between the processing of active components is crucial. This important feature distinguishes LARA++ from active router implementations such as CANEs [ ], Router Plugins [ ] and Scout [ ].

Packets requiring reclassification are separated from packets that are awaiting initial classification, so that the classifier can process packets waiting on the internal queue in preference to those on the external queue and thus minimise packet latency.

Packets being processed by a LARA++ node are given a packet context which is used to store state information, such as its progress through the classifier. On arrival, the user flow of the packet is calculated and a flow key is generated to facilitate flow filter lookups. This key is stored in the packet context. When checking flow filters, the key is used to perform a lookup in the hash table for the node. This operation yields a shortlist of candidate flow filters, which are then checked individually.

In each node in the classification graph, filters are processed starting with flow filters, then general filters, graph filters and finally the default graph. Graph filters are processed last in a node to ensure that all active components are processed before progressing to the subsequent node in the classification node. A classification is made if the patterns of any filter match the packet. If the match is made with an active component filter, the packet is sent to the associated component for processing, and resumes the classification process at the next filter in the same classification node on its return. If the match is made with a graph filter, or with the default graph filter, which is matched if no other graph filter could be applied, classification terminates in the filter input node and resumes at the start of the filter output node.

## Filter Processing

The creation of a service composite for each packet is based upon the packet filters. Section introduced the notion of filter patterns. Each filter type, general, flow and graph filter, contains such a pattern as one of its attributes. While the expression of packet filters in this way is convenient and extremely flexible, it comes with an inherent overhead. The position of fields that might be identified by the filter pattern (e.g. **TCP_HEADER**) can change from packet to packet due to extra headers and options. This means that the absolute offset must be recalculated for each packet. The impact of the operation can be somewhat lessened if the classifier maintains a list of packet characteristics (e.g. protocol headers) that have been identified during the classification of the packet in its journey through the active router. These *features* can then be used in the offset calculation, rather than having to parse the packet to locate these features each time they are required. For example, the classifier could store the offset of the IPv header in the packet when the header is encountered so that subsequent filters can use it in offset calculations. Because of this approach, it is not a coincidence that most headers have one or more dedicated classification nodes in the classification graph this is a property of the composition model

In order to facilitate this optimisation graph filters are given an additional property known as the *focus translation* The packet context contains a stack of foci and the packet begins classification at the first classification node with a single focus of zero If a graph filter is matched or default graph is encountered a new focus is pushed on the stack The new focus increases decreases the previous offset by the focus transla tion of the matched graph filter or default graph For example the focus translation of a graph filter branching between an IP header and a TCP header would be the size in bytes of the IP header A focus that pointed to the start of the IP header would point to the start of the TCP header subsequent to the processing of the graph filter

In order to find a feature of the packet for use in offset calculation the problem is reduced to one of searching for the desired feature identified by the classification node on the stack of foci The focus stack model was chosen because it is likely that most attempts to examine features of packets will be made closest to the focus of the packet in the current classification node Since the most recent foci are placed at the top of the stack searches for foci usually find a match within a few attempts For ex ample a filter identifying the protocol field in the IP header will normally be placed in the IP header classification node thus the operation to locate that field i e "Focus {IP_HEADER}+IP_PROTOCOL" will find the target focus at the top of the stack

Another potentially heavyweight task in filter processing is the computation in volved in calculating the offset of packet fields The fact that packet features are not always of constant length e g IP options and padding can cause the length of an IP header to vary creates a need for flexible expressions such as the one above in order to specify the focus translation for graph and default filters Since the majority of network traffic can be categorised into just a few payload types many filter patterns and graph filters will need to be checked against every packet passing though the node Given the frequency of the evaluation of offsets and the fact that packet filters do not change after filter installation it is best to move the overhead of evaluating the semantics of the expression to the installation time of the filter We use a just in time compiler that translates the safe machine independent expressions into native ma chine code at the time of filter installation Consequently execution of the compiled expressions is very lightweight only a few CPU cycles This allows focus transla tions and packet offsets to be calculated efficiently and flexibly on a per packet basis

## Performance Measurements

On completion of the prototype classifier we took measurements of the throughput under three different scenarios These scenarios were intended to lend proof of con cept to the LARA model for flexible and dynamic composition Each of the sce narios operated over the same populated classification graph albeit with different processing characteristics for each one

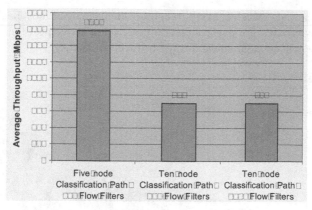

**Fig.** Classifier Throughput for Pre-defined Packet Paths. These tests were performed on an Athlon XP with Mb RAM running Windows using simulated packets, thus producing a reflection of the Classifier's capabilities untainted by other active routing activities.

Experiment one involved a type of packets chosen so that the traffic passes through classification nodes in the classification graph. The packets were checked against general filters and flow filters. The second experiment used a packet content that causes the traffic to pass through classification nodes in the classification graph. The number of general filters was doubled in order to impose roughly the same processing load per node as in test one, whereas the number of flow filters on the path was kept constant at. By comparing the throughput of the first and second tests, we expected to find the processing load to be proportional to the number of classification nodes through which the packet travelled. The third scenario involved the same packet format and number of general filters on the packet path as the second test, but the number of flow filters on the classification path was doubled. The objective of this experiment was to confirm that adding extra flow filters does not proportionally decrease performance, all other things remaining equal.

Figure presents the results of these three experiments calculated over million packets. As expected the throughput roughly halved between experiment one and two because the number of classification nodes on the packet path doubled. The results show that the graph filters and general filters do not scale well. Fortunately, their numbers are not related to the number of users of the active router, unprivileged users can only install flow filters, and hence do not have to scale to large quantities. By increasing the number of users of the active network, mainly the number of flow filters will increase proportionally. Between experiment two and three the number of flow filters doubled, but performance was virtually unaffected. With an average drop in throughput of between these experiments, we have shown that the use of flow filters allows the classification model to scale well as the number of users rises.

Further experiments were performed with the aim of measuring the packet latency. Figure illustrates a breakdown of the time taken to perform different stages of classification. Six states of processing have been selected to represent the complete passage of a packet through the classifier, and the figure shows how these states account for the total latency of a packet. The average latency imposed by the classifier on an individual packet is ... μs which is a tiny fraction of the latency of most packets travelling though a passive router. For packets that do not require active processing, cut through paths in the classification graph can further reduce that latency.

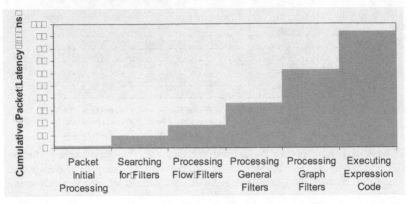

**Fig. ...** Breakdown of the stages of packet classification, measured in tenths of microseconds. The classification graph and its population were chosen such that ... flow, ... graph and ... general filters were checked. Of these ... flow, ... general and ... graph filters were matched.

Of the total latency, more than half is accounted for by the complexity of the classification graph. The complexity of the path through the classification graph undoubtedly has a direct impact upon the latency. Thus, the main determinant of the packet latency is the complexity of the packet itself. The majority of packets have only a MAC header, a network header, a transport header and a payload but rarely have many options. They would therefore take a "shortcut" through the classification graph, avoiding the extra classification nodes required to process these optional headers.

The average latency of the classification stage processing flow filters is comparatively small (~ μs) in the context of the total packet latency. The measurements described above show that doubling the number of flow filters in the classification path reduces the throughput by less than ... The impact of such a proportion added to the latency of this classification stage would barely be noticeable. The classification latency of packets is therefore largely unaffected by a change in the number of flow filters installed on an active router.

## ...Related Work

A common objective of most active network approaches is to expedite network evolution through solutions that enable extensibility of network functionality by way of dynamically loaded code. Most active network approaches, such as ANTS [ ], NetScript [ ], PLANet [ ], SmartPackets [ ], accomplish this through software plug ins or a similar form of active code integration. The limitations of such plug in based approaches to extensibility have been revealed in a previous study [ ]. The remainder of this section compares further, more closely related approaches to LARA...

The CANEs execution environment [ ] implemented on top of the Bowman No-deOS provides a composition framework for active services based on the selection and customisation of a generic "underlying program". The underlying program can be tailored for a type of packets or set of streams by injecting customised code into well-defined slots in the program. The packet filter mechanism selecting the underlying program can be configured to match arbitrary patterns in the packet. This flexible classification approach allows Bowman to dynamically deploy new protocols at run-time, like LARA++. However in contrast to LARA++, Bowman has a number of re-strictions. First, Bowman restricts classification to the selection of an underlying pro-gram, i.e. once an appropriate underlying program has been identified, the service composite is fixed and only dependent on the plug-ins. Second, although Bowman appears to allow multiple underlying programs to be selected, the literature implies that only a copy of the packet can be sent to each logical input channel which prevents implicit active program co-operation. CANEs further restricts service composition. The static nature of the underlying program for any given execution environment is naturally inflexible. One needs to make assumptions about the customisable aspects of the program at instantiation time of the execution environment.

The Router Plugins architecture [ ] also uses a plug-in based composition model, whereby an underlying data structure or program defines the "glue" for the service composites. The fact that these composition structures are defined at compile time of the kernel limits extensibility to predefined *gates*. LARA++, by comparison, allows the dynamic extension of the classification graph, i.e. allows the creation of new clas-sification nodes at run-time, and also overcomes the limitations that only one plug-in can be incorporated per gate, i.e. many active components can be inserted per node [ ].

Further related works are the modular router architecture Click [ ] and the config-urable operating system Scout [ ]. Both use a graph based composition approach like LARA++ to support extensibility of the communication subsystem through so called modules. However, since configurability in both cases is limited to the compile-time of the system, dynamic introduction of new services is not possible. This analysis shows that providing the service composite for plug-in or component-based solutions should be extensible at run-time. Assuming an underlying graph structure or program that cannot be dynamically changed is not necessarily suitable for the lifetime of the system, and thus, limits extensibility unnecessarily.

## Conclusions

In this paper we have presented a novel framework for managing the creation of serv-ice composites in an active router. The LARA++ composition framework supports dynamic integration of router extensions (active components) at runtime. The classifi-cation based service composition model enables flexible integration of extended router functionality at any point in the packet processing path. A classification graph, representing the packet processing path on the router, provides the necessary man-agement structure for the integration of the software extensions. The use of packet filters as a means of binding the software components allows the composition mecha-nism to dynamically incorporate new functionality at run time.

The service composition model is sufficiently flexible to allow the creation of a service composite for each packet passing through the router. Using packet filters to compose active services also has the advantage that active computation can be applied transparently. The application of an active extension is based on the packet content, i.e. any bit pattern can be used to trigger the processing of an active component.

The LARA++ composition model is capable of managing both competition and co-operation between users of an active router. It allows unrelated users to partake in the programming process of the active node in a structured fashion. The model provides sufficient semantics to structure independent software extensions in a meaningful way.

Finally, we have presented the evaluation results of the LARA++ composition framework. The results show that the processing latency imposed by the classifier on an individual packet is less than one hundredth of a millisecond, which is a tiny fraction of the latency introduced by a normal edge router. This shows that the inclusion of the LARA++ composition framework has a negligible impact on the overall latency of the packets passing such a node. The results also demonstrate that the introduction of the flow filters allows the composition model to scale exceptionally well in terms of both the throughput of the active router and the latency of packets being routed, and does so without the number of users significantly reducing performance.

# References

1. R. Cardoe et al.: "LARA: A Prototype System for Supporting High Performance Active Networking". In Proc. of IWAN '99, June 1999.
2. D. Decasper, Z. Dittia, G. Parulkar, B. Plattner: "Router Plug-ins: A Software Architecture for Next Generation Routers". In Proc. of SIGCOMM, pages 229-240, September 1998.
3. D. J. Wetherall, J. V. Guttag and D. L. Tennenhouse: "ANTS: A Toolkit for Building and Dynamically Deploying Network Protocols". In Proc. of OPENARCH, April 1998.
4. M. W. Hicks and J. T. Moore and D. S. Alexander and C. A. Gunter and S. Nettles: "PLANet: An Active Internetwork". In Proc. of IEEE INFOCOM '99, 1999.
5. B. Schwartz et al.: "Smart Packets for Active Networks". In Proc. of OPENARCH, 1999.
6. S. Schmid, J. Finney, A. C. Scott, W. D. Shepherd: "Component-based Active Network Architecture". In Proc. of IEEE Symposium on Computers and Communications, July 2001.
7. K. Egevang et al.: "The IP Network Address Translator (NAT)". RFC 1631, May 1994.
8. S. Merugu et al.: "Bowman and CANEs: Implementation of an Active Network". In Proc. of 37th Conference on Communication, Control and Computing, September 1999.
9. Y. Yemini and S. da Silva: "Towards Programmable Networks". In Proc. of IFIP/IEEE International Workshop on Distributed Systems, Operations and Management, October 1996.
10. M. W. Hicks and S. Nettles: "Active Networking Means Evolution (or Enhanced Extensibility) Required". In Proc. of IWAN 2000, October 2000.
11. A. Montz et Al.: "Scout: A Communications-Oriented Operating System". In Operating Systems Design and Implementation, pages 200-210, 1994.
12. R. Morris, E. Kohler, J. Jannotti, M. Kaashoek: "The Click Modular Router". In Proc. of ACM Symposium on Operating Systems Principles, pages 217-231, December 1999.

# Author Index

# Lecture Notes in Computer Science

For information about Vols. 1–2452

please contact your bookseller or Springer-Verlag